LITERARY KNOWLEDGE

Also by Paisley Livingston:

Ingmar Bergman and the Rituals of Art

LITERARY KNOWLEDGE

Humanistic Inquiry and the Philosophy of Science

Paisley Livingston

Cornell University Press

ITHACA AND LONDON

First published 1988 by Cornell University Press.

International Standard Book Number 0-8014-2110-1 (cloth)
International Standard Book Number 0-8014-9422-2 (paper)
Library of Congress Catalog Card Number 87-47821
Printed in the United States of America
Librarians: Library of Congress cataloging information appears on the last page of the book.

The paper in this book is acid-free and meets the guidelines for permanence and durability of the Committee on Production Guidelines for Book Longevity of the Council on Library Resources.

FOR METTE

Contents

Acknowledgments ix
Introduction: Literary Knowledge, or the Shadows of Science 1

PART ONE: ON LITERARY THEORY

1 From Prophecy to Inquiry 9

PART TWO: OF SCIENCE AND NATURE

2 In Defense of Epistemology 38
3 On Scientific Realism 80

PART THREE: THEORY AND THE QUESTION OF THE HUMAN SCIENCES

4 Idealism and Naturalism 121
5 Arguments on the Unity of Science 147

PART FOUR: LITERARY KNOWLEDGE

6 Literary Explanations 200
7 *Hypotheses Fingo* 243

Index 269

Acknowledgments

I am indebted to David Williams, chairman of the Department of English at McGill University, and to my other colleagues in the department, for granting me the freedom to develop my teaching and research in a variety of directions. This kind of freedom is, I think, all too rare in our world of disciplinary barriers and departmental priorities. I also acknowledge the Social Sciences and Humanities Research Council of Canada, for a research and leave grant in support of this project, as well as McGill University, for granting me a badly needed sabbatical year. My thanks were also earned by Cornell University Press's three readers, whose generous comments have helped to improve the manuscript, and to Cornell's editors as well, especially to John Thomas for his worthwhile corrections.

One often sees acknowledgments in which a spouse is thanked for her patience, support, diligent proofreading, and so on, and I have every reason to repeat similar words of thanks. Yet as I am fortunately able to share every aspect of my work with my wife, Anne Mette Hjort, her place in these acknowledgments by far surpasses the norm: thus I dedicate this book to Mette, whose knowledge, incisive comments, and general influence on me have had everything to do with whatever may be worthwhile in this volume.

Finally, I thank my father, who showed me at an early age how it is that airplanes can fly.

PAISLEY LIVINGSTON

Paris

LITERARY KNOWLEDGE

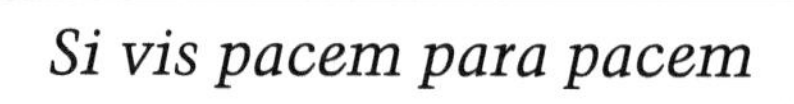

Si vis pacem para pacem

INTRODUCTION

Literary Knowledge, or the Shadows of Science

To speak of the shadows of science is to call for a clarification. What is it, in fact, that needs to be clarified? Does science really have any shadows? Perhaps it will be a matter here of the dark side of Enlightenment, of Newton's alchemy, of the lurid glow of napalm, and science's other clouds. Or do we want to say that only a bright light can throw a shadow, that what needs to be illuminated are the vestiges of darkness lurking in the lantern's path? Will we choose geometry or finesse, mastery or flights?

And what is the role of literature here? Is literature—so much make-believe and dreaming—to be cast as the mere shadow of science, with its reliable plans and efficient machines? Or is literature to play the prophet, protesting to the point of madness against the murderous weapons of science? Will it be the leap of Empedocles into the volcano or the voice of the *raisonneur*, enchantment or experiment? Where is literature to be placed in the great chiaroscuro of the West?

It will indeed be a matter, in what follows, of the Enlightenment and its contraries, but if we have learned anything, it is that the dichotomies drawn in such schemes are far too tidy to be trusted. Only in someone's puppet show is science paraded as all light and fairness, without its insanities, destructions, frantic growth, and pretending. Only in someone's narrow device is literature only hearsay and decoration, lacking all rigor and truth. But this does not mean that science has no brilliance, and literature no folly. If there is a great divide to be traced, then, it is not one that cleanly sunders knowledge from fiction, the hard from the soft, the exact from the fuzzy; it is an opposition that concerns them both, for in science *and* in literature we find hypothesis and fact, creation and discovery, the delicate and the correct. Scientists write and invent, and men and women of letters make subtle models of the patterns that connect and explain. All of them make mistakes, some glimpse facets of the truth.

We need not read fiction to learn of the comets and stars, nor should we go to a performance of Molière for a cure. Must we now study science to enjoy our poems and fictions, and to create new ones? Not at all. There is no talk in what follows of effacing all such differences. But when it is a question of that hybrid figure, the contemporary literary scholar, the issues are more complex. This figure may be mistaken for the leisured elucidator, devoted to the pastime and refinements of art, but such an image is contradicted by the literary connoisseur's presence in the modern university. Is the critic a kind of curator, then, a walking memory, guardian of one time, one place, one life and works? But we have books for that, and not even a mountain of facts makes a piece of real research. The professor of literature must do research, even if its nature and aims are unclear, even if this research may have little to do with the needs and interests of the students, even if it is largely a myth that the results of this research are addressed to the reading public at large—whose appreciation of the classics is to be enhanced.

Once literature is chained to the business and career of knowledge, the rules of the game change. This conjunction does not go without saying, it is not the least bit obvious, but it is today a matter of fact. A simple answer to the question concerning the role of theory in the literary disciplines has to do precisely with this fact: in all of its various forms, theory is an effort to come to terms with what is incongruous and complex in the very notion of a literary knowledge, which concerns the knowledge *of* literature as well as a knowledge *in* literature. Theory, then, is about criticism and what critics should and should not try to do with the primary works. By the same stroke, theory is necessarily about these works as well, and what they can and cannot be expected to offer us. In this sense, literary theory is not a newcomer to the discipline of literary analysis, it is a direct and necessary consequence of the very existence of such a thing as a literary discipline with its implicit claim to a literary knowledge. As long as this discipline is around, or at least as long as it is taken as a serious intellectual enterprise among other disciplines and inquiries, René Wellek's phrase will remain as apt as it was when he first wrote it: "Literary theory, an *organon* of methods, is the great need of literary scholarship today."[1]

Literary theory, one may protest, is already an industry, and it would not be out of place to wonder what another volume may be expected to offer to a field that at times appears to be saturated. Are there really any breakthroughs to be expected in second-order talk about how critics talk—or worse still, in theorizing about theorists? Have not all the positions been covered? It is certainly true that literary theory has made great progress over the last decade; gone are the days when it was possible to brush aside the conceptual problems of criticism and its methods in a genial and erudite chat.[2] Anyone who wants to take the issue of literary knowledge seriously must come to terms with the range of established positions on the subject.

I write this volume as a latecomer who entered the field at a time when theory was already well on its way to being an established part of the literary disciplines. I have been able to profit immensely from the intellectual and political contributions that theory-oriented critics have made in the past two decades. Thus, if this book challenges some of the positions taken by a large number of these critics, my criticisms must not be taken as a call for a return to some previous understanding of literary research. On the contrary. If what I am saying must be assimilated to one of the recognized positions, then I would prefer to be mistaken for one of those who complain that the true promises of theory are betrayed unless we make a radical break with traditional critical practices and assumptions.[3]

In other words, I believe that a rigorous skepticism is wholly appropriate in regard to the knowledge claims implicit in the largest part of what may be called "business-as-usual" literary interpretations, and thus one does not find here any apologies for so-called practical criticism. However, I do not agree that this skepticism is appropriate as a general attitude or that one may take the literary example as a basis for challenging *all* knowledge claims. I do not agree that critics should at this point strike over the words "knowledge," "reasons," "evidence," "truth," and, of course, the unspeakable "science," and then ask themselves what radically different

[1]René Wellek and Austin Warren, *Theory of Literature* (New York: Harcourt, Brace & World, 1956), 19.

[2]An example is Harry Levin's *Why Literary Criticism Is Not an Exact Science* (Cambridge, Mass.: Harvard University Press, 1967).

[3]Thus I am in agreement with Rodolphe Gasché's views on the continuity between a certain deconstructive style and New Criticism. See his "Deconstruction as Criticism," *Glyph*, 6 (1979), 177–216.

strategies, games, pleasures, diversions, and subversions may be devised. A coherent skepticism does not warrant any such moves, nor is its much-touted rigor really so unassailable. Of course the word "skepticism" hardly embraces the diversity of views at stake, but it may nonetheless usefully serve as a first approximation of the very general point of my divergence from theoretical work that I otherwise find to be of value.

It is in regard to the skeptical tendencies of contemporary critical theory that the issue of the natural sciences comes into the foreground. Although I was at one point an avid student of chemistry, I decided quite some time ago that my interests lay, not in the mysteries of molecules, atoms, and beyond, but in those of the world of humanity. Thus it is not as the result of some scientist's *parti pris* that I am dismayed by the absurd statements about science I have heard, on a regular basis, in the literary seminar room. These statements range from subtle but perverse argumentations to the rather blunt and pathetic "science is just rhetoric too," but the thrust is always the same: the poetic and critical luminaries could not justifiably be thought to be shadowy in relation to scientific work because the latter is all darkness anyway; if progress in literary criticism is a dubious matter, so is that of science—and so on. I think these views are definitely ill informed and probably dead wrong, and yet I have noticed that they are repeatedly asserted in literary circles in a self-assured and dogmatic manner. Nor is this only a matter of cursing an overly privileged neighbor, for critics actually orient their own research in relation to these same dubious ideas about science and knowledge. It is for this reason that a large part of this book is engaged in challenging this dogmatism, my goal being to reopen literary criticism's dialogue with the natural and social scientific disciplines. My conviction is that literary theory should seek to clarify the relations between literary research and other models of knowledge. I contend that literary critics should recognize that the basic scientific project is not reducible to scientism, or to a group of arbitrary worldviews—as the sterile opposition between positivists and relativists would have it.

In the first part of this book, I discuss the role of theory in the literary disciplines and survey some of the major ways in which theory may be defined. I argue that theory should play an epistemological role and that it should contribute to a better understanding of the relations between literary research and the scientific model of knowledge.

In the second part of this book, the epistemological role of critical theory is defended against fashionable doubts about the viability of epistemology. Different types of skepticism are distinguished, and the question of the justification of belief is shown to be irreducible to the kind of institutional analyses that have characterized the "sociological turn" in the philosophy of science and in literary theory. Given this basis, the work turns to examine the question of the status of natural scientific knowledge and mounts a series of arguments against a cluster of views designated as "framework relativism." Literary critics are shown that it is dogmatic to draw conclusions about science on the basis of the writings of Thomas Kuhn and Paul Feyerabend, and recent arguments in favor of scientific realism are presented. More specifically, I isolate from among the various forms of realism the nonreductionist version that has recently been defended in a clear and rigorous manner by Richard N. Boyd, and I present some of the main arguments between this realist position and the constructivist and empiricist alternatives. Although I make no claim to present anything approaching a comprehensive and conclusive viewpoint on the philosophy of science in this section of the book, I do establish the inadequacy of the assertive forms of relativism and skepticism that one frequently encounters in literary theory.

In the third part of this book I criticize a variety of epistemological arguments against the viability of extending the natural scientific model of knowledge into the domains of history, culture, and society. A discussion of Edmund Husserl's later work isolates the idealist assumptions backing certain of these arguments, and reference to Roman Ingarden reveals their pertinence to literary theory. Ludwig Wittgenstein is identified as the source of another major set of claims supporting a fundamental dualism of methods. I show that Wittgenstein's work lends no real support to the framework relativism urged by the postmodernists and go on to criticize a form of argument typically employed by that camp. Thus I criticize what I call the "always already" argument, which is used to erect an a priori barrier against certain forms of inquiry into human and cultural realities.

Some critics may grant my rather basic point about science and may prefer to know what is supposed to follow, in more concrete and positive terms, from a moderate recognition of the achievements and progress of knowledge within the natural sciences. Tentative and partial answers to that major question are explored in the final part, where I take up what I characterize as the central debate in literary theory—the question of the validity of interpretation—and analyze the aesthetic and communicational assumptions that have drawn this debate into a sterile impasse. A discussion of models of scientific explanation proposed in the philosophy of science provides a useful background for an examination of literary-critical research, an examination that suggests that many critical inquiries falter as a result of a basic problem of orientation. I distinguish between different lines of inquiry in literary research and show how a typical variety of "megaphone criticism" confuses these distinct tasks. Two alternative avenues of literary inquiry are discussed, one historical and sociological, the other heuristic and interdisciplinary. The goal of these analyses is, finally, a defense of literary knowledge, as well as a clarification of its interdisciplinary relations.

As a conclusion to this introduction, here are two of my most basic starting points:

1. Construed broadly, the literary canon comprises a vast and extraordinarily rich body of symbolic artifacts. This richness has to do, not with the movements of the comets, but with the complexities and conditions of human experience. Is it not strange and wrong that the overarching tendency within our societies is to view these materials as of marginal utility—and, particularly, as of little cognitive value? (The due respect paid to the mausoleum is another matter.)
2. Literary scholars have devoted careers to the study of these documents and are on the whole good and well-informed readers. Yet the fruits of their knowledge—its recorded results, for example, in the form of literary publications—remain marginalized—not only in relation to social life as a whole (which is not surprising in a barbaric environment), but even in relation to the main streams of research in the human sciences—in psychology, anthropology, economics, political science, and so on. Is this not strange and wrong? Is it not stranger still that literary scholars are so infrequently capable of saying why this should not be so? Can they do nothing about it? It may be true, as the poet said, that "il mondo è fango" (the world is mud)—but whoever said critics could keep their feet clean?

PART ONE

ON LITERARY THEORY

CHAPTER I

From Prophecy to Inquiry

To speak of critical theory and of literary theory, as is increasingly common in literary circles, is anything but precise. Nor is it only in relation to some highly idealized standard of precision that this vagueness can be measured. On the contrary, there is a profound and thorough confusion about what is at stake in talk of theory within the literary disciplines, a confusion that surpasses the measure of vagueness that may be expected—and readily accommodated—within the terminology of any discipline. Nothing is obvious, nothing can be taken for granted, nothing is a matter of consensus, whenever there is a question about the nature and role of theory in the literary fields.

Is there any such thing as literary theory? Even on this most basic issue opinion is divided. Some would hold that theory is a necessary part of every literary discipline. Although theory has become a predominant and highly visible institutional fact only over the past two decades (it first appears as a special rubric in *The Year's Work in English Studies* in 1981),[1] one may argue that, in a correct understanding of the notion, theory has always been present. Theory is "already always" there (*schon immer*, as Husserl frequently wrote). Those who raise their voices against it, calling for a return to some kind of direct or practical form of criticism, can easily be shown to be urging one form of theory—the antitheoretical theory.[2] Yet to argue

[1]Robert Young, "Literary Theory," in *The Year's Work in English Studies*, ed. Laurel Blake, 62 (1981), 7–67.

[2]See, for example, the controversy in *Critical Inquiry*: Steven Knapp and Walter

the opposite, to point to the strict inappropriateness of theory within literary studies (and more generally, in the humanities), one need only take up a different and equally legitimate set of meanings associated with the term. Thus, if by "theory" is meant "deterministic automaton modeling a closed set of factual statements about the literary domain of reference," it is clear that no literary theory presently exists and that none is likely to appear. It is therefore crucial to say in what sense "theory" is to be taken, for in the absence of even a partial and temporary clarification along these lines, all debates over the place of theory in the discipline will remain a *dialogue de sourds*. What follows in this chapter is an attempt to contribute to such a clarification. I first identify several major ways of construing the term, and with it, the proper role of theorizing in literary research. My approach to this task is more conceptual than descriptive and exegetical. For reasons that become obvious in a moment, I think that the last thing needed is another book organized in function of the names of prominent critical theorists, taken up in a chronological or historical order. But this does not mean that I pretend to stand outside the literature on theory and the ongoing debates it manifests. Moreover, the goal of my discussion of alternative notions of theory is not merely to construct a tidy *divisio* but to take a clear stance on what kind of theory is most appropriate and useful. The analyses that follow in the rest of this book are presented not as the completion of such a theory (the very idea being an absurdity in my terms) but as a first contribution along these lines.[3]

Substantive and Invisible Theory

Theory, it seems, was once a sacred matter. We know that the Greek *theoria* was used to refer to a group of official envoys sent to represent a city-state on the occasion of religious festivals or games (an Aeschylus fragment bears the title *Theoroi e Isthmiastai*), and the verb form designated the fulfillment of the ritual role played by these sacred onlookers. Although seemingly profane uses referred more simply to seeing, observing, and being a spectator, it has been

Benn Michaels, "Against Theory," *Critical Inquiry*, 8 (1982), 723–42; the articles in response, *Critical Inquiry*, 9 (1983), 726–89; and the antitheory theorists' reply, 790–800.

[3]An excellent source of critics' attitudes on theory is the survey "Literary Theory in the University," *New Literary History*, 14 (1983), 411–51.

argued that the religious connotations were never very far away.[4] Is theory a matter of religion, then, and do the theorists form a kind of priesthood? Etymologies—even the most factual and rigorous—may tell us very little about present usage, nor should they be taken as a way of uncovering a deeper and more essential meaning, that of the "origin." Not only are these dubious notions, but it would have to be shown in the first place that the same words are being related. Moreover, even if we wished to take the ancient usage as a kind of guide, there would remain the immense difficulty of understanding the Greek practices with which these words were associated. What was at stake in these games and what role did the *theoroi* play in them? In the absence of answers to such questions, any analogies based upon ancient usage are spurious.

Yet the etymology is suggestive of one aspect of theory, as there is a sense in which theoretical activities are to some critics what the priest is to the cult. It is also true that some theorists actively play the role of the prophet.[5] As preeminent interpreters of the literary oracles, theorists play the role of charismatic leaders, models whose prestigious way of thought and life is believed to be profoundly different from that of the everyday critic. Theory, then, is a model or style to be adopted by others; it is a prophecy about the future of criticism, a prophecy realized as often as there are other critics who follow in the theorist's path. What is wrong with this approach to theory is not the venality of the priest—a venality that in our day is so exacerbated as to vitiate the analogy, so that one would have to speak of sacred prostitutes instead of priests. Rather, the cult goes wrong because it is not a theorist's exemplary life and mythical status that should come first, but the validity and value of the work. It is a distortion to imagine that theories and criticism can only refer to those who create them, a distortion perpetuated in the literature whenever author-studious gestures supplant arguments and evidence. Even if we were to grant the wild assumption that criticism and theory should be read as literature, we would still need to know something about how literature should be read, and it has not been

[4]Hannelore Rausch, *Theoria: Von ihrer sakralen zur philosophischen Bedeutung* (Munich: Wilhelm Fink, 1982).

[5]A similar analogy is carried to great length by Claus Borgeest, "Über die Funktion der Kunst," in *Literatur und Kunst—Wozu?* ed. Siegfried J. Schmidt (Heidelberg: Carl Winter, 1982), 29–54. In a North American literary-critical context, the antipriesthood stance is voiced by Jonathan Culler, "Comparative Literature and the Pieties," in *Profession 86*, ed. Phyllis Franklin (New York: Modern Language Association, 1986), 30–32.

established that conducting author studies is the best way to find out. For these and related reasons, what I discuss in the following is, on the whole, theories, not theorists.

In referring above to a *substantive* type of theory, I wish to signify a broad class of critical practices and attitudes which manifest a particular approach to the nature and role of critical theory. In a sense, this approach is close to the priestly type insofar as a model exercizes an immediate and powerful influence upon the critic's particular judgments. Although theory could be understood to involve a kind of metacritical and methodological reflection, in the substantive approach theory provides a set of themes that are directly imputed or applied to the primary material. Theory in this sense is anything but a method of doing criticism, for to speak of a method is to imply the possibility of a kind of partially rational choice between alternative ways of doing something. Yet the reader who is guided by a substantive theory already has drawn from it a set of notions and habits which largely determines the particular cases to be selected for commentary and what is to be said about them. The mediocre versions of any of the major approaches to the study of literature exemplify this shortcoming, for be it a matter of psychoanalysis, semiotics, Jungianism, or deconstruction, the text is the dough and the theory is the cookie cutter. But cookie-cutter criticism need not only be the effect of a clumsy application, for there are theorists and theories of a truly substantive bent. Whenever a theoretical text organizes within itself a body of general statements concerning the universal nature and essence of literature, of meaning, of the human psyche, of culture, of history, whenever it advances an exemplary and global model of judgment and of understanding, only a *subversive* reading, one that brushes against the theory's own grain, can prevent the theory from providing another cookie cutter. A substantive theory, then, can be defined as one proposing a final answer to the question of the nature of literature and of criticism, an answer immediately applicable to the case at hand in the form of a set of privileged themes and generalities. It could be objected here that only the purest of skepticisms are wholly devoid of substantive theses, but the remark is hardly pertinent in relation to the kinds of simplistic thematic reifications that abound in literary theory. Indeed, the literary skeptic's favorite theme—"the impossibility of a valid reading"—is simply another device by means of which theoretical and critical inquiry are reduced to redundant applications of preestablished "truths."

This substantive type of theory is what makes theory in general anathema to more traditional critics who believe in the priority of detailed and specialized investigations and shudder at the sweeping generalities uttered by theorists. These shudders are often justified, but this is not to say that everything is as it should be within the world of particularist research. For the latter often proves to be the domain of what I would call "invisible theory." In fact, the substantive and invisible forms of theory mirror each other as complementary forms of dogmatism. In both cases, the process of hypothesis formation essential to all genuine inquiry is lacking. On the one hand, it is skipped in favor of a pseudoempirical and "immediate" approach to the "particular facts." But such an approach can only be as good as the implicit and general assumptions that in fact guide and organize the inquiry. Again and again, the supposedly nontheoretical approach amounts to a tacit reliance upon a complex host of invisible theories: the sedimented and unexamined theory of genres, a prejudicial nationalistic parceling out of "literatures," an unreflective periodization, a Eurocentric and elitist canon mirroring a "great man" view of history, a wholly idealist aesthetics, an arcane and incoherent semantics, colonial ethics, and so on. On the other hand, there is substantive theory, which eschews the formation of hypotheses in favor of pronouncing generalizations that are immediately projected as the accurate representation of the domain in question. Confronted with the false choice between the two, we should do our best to avoid condoning either folly; literature is not to be mistaken for an illustration of the new Professor X's theory, nor is it best viewed as a jumble of disparate facts rumbling about in the dusty drawers of the traditional categories.

Formal Theory

Literary scholars should become more aware of the fact that in other disciplines the term "theory" is sometimes construed in a manner that differs quite fundamentally from their own ways of using the term. This is not to propose that literary scholars look to these other fields for the true or correct definition of "theory," but to note the heuristic value of such comparisons, which may contribute to a clarification and strengthening of the literary usage. The notion of "formalized theory" is a case in point, for what the term designates is an understanding of theory which, if it has an important role in other fields, cannot be said to be prevalent in literary research.

What would it mean, for example, for literary studies to adopt the definition of "theory" that has been stated quite succinctly in the following manner? "A theory is a systematically related set of statements, including some lawlike generalizations, that is empirically testable."[6] Presumably, if one were to use this formula in a definition of literary theory, it would simply be necessary to add that the statements figuring in the theory be about literature. In order to highlight what is particularly interesting about such a proposal, we may contrast this first definition with one given in a recent work surveying the history of theories of drama: "The term 'theory,' then, I have taken to mean statements of general principles regarding the methods, aims, functions, and characteristics of this particular art form."[7] The contrast is striking. Although both definitions stress the general nature of the statements of a theory, the second lacks any clause pertaining to the veracity, falsehood, or testability of these statements. This is a far-reaching difference, one I discuss at some length in the final two chapters. A second major difference pertains to the stipulation, in the first definition, that the statements be systematically related. It is important not to be led astray here by the loose sense in which the word "system" is often employed in literary circles, a sense that is inappropriate in regard to the understanding of theory exemplified by the first definition. In Richard Rudner's notion of theory, which is in no way idiosyncratic, the concept of "system" is not synonymous with general notions of "orderliness" or of "fitting together." Rather, this concept imposes a rigorous set of formal requirements upon the builder of a theory. Rudner goes on to equate his notion of a systematic relationship between the statements in a theory to the concept of "formalized language." The latter term designates a set of elements or terms and the transformation rules that determine the class of well-formed strings of elements that can be generated. There must be a decision procedure whereby every string of elements can be determined to be either well formed or not; moreover, the well-formed strings are either axioms or theorems of the system. Another way to put this is to say that a formalized theory is one ordered by a logical syntax; one may also speak of this kind of system as a calculus, the statements of which could be computed by a Turing machine.[8] Such a theory has a factual component insofar as

[6]Richard S. Rudner, *Philosophy of Social Science* (Englewood Cliffs, N.J.: Prentice-Hall, 1966), 10. Robert K. Merton construes theory similarly in his *Social Theory and Social Structure* (Glencoe, Ill.: Free Press, 1949), 90.

[7]Marvin Carlson, *Theories of the Theatre: A Historical and Critical Survey, from the Greeks to the Present* (Ithaca: Cornell University Press, 1984), 9.

[8]For a broad discussion of related notions of calculation, see Heinz von Foerster, "Disorder/Order: Discovery or Invention?" in *Disorder and Order: Proceedings of the Stanford International Symposium, Stanford Literature Studies* 1, ed. Paisley Living-

it is interpreted, that is, only when its terms are given semantic relations specifying what its elements designate (thus Rudner refers to "empirical testability").

It should be clear that we do not presently possess any literary theories that fulfill such requirements. The most "formalist" literary theories are but quasi-formalized discussions, for the conceptual schemas they present usually amount to broad typological frameworks, attempts to define series of terms, and classificatory schemes that pullulate with nonordered sets and exceptions. Although frequent references may be made to codes, transformation rules, grammars, and such, these remain purely hypothetical postulates unsupported by a rigorously formalized calculus, and in the absence of such support, their status may be likened to that of a blank check. In this regard, a lot of the talk of "rigor" that one encounters in literary theory is misleading, and one may wonder whether the word is sometimes exploited for its technical connotations. To say that *x*'s essay on *y* was a rigorous analysis usually means that *x* said something incisive, that the essay was detailed, careful, accurate, and so on; it does not mean that *x* wrote an algorithm for *y*. Yet there are a lot of very good reasons that no formal rigor has ever been achieved in literary debates and that it is not a good idea to set out immediately to fabricate a theory—in Rudner's sense—of the literary domain. No *plaidoyer* for any such project is being made here. The reasons, I have said, are numerous. For one, we have not yet established anything about what constitutes the domain of literary facts to be modeled by the sentences or strings of our literary system. In the absence of any such delimitation of the domain of reference, in the absence of a correct decision about the nature of the basic units of analysis, it is wholly unclear what any tinkering with formal tools might achieve. Nor is it obvious to many critics why one would want to construct any such theory in the first place. What is its supposed utility and interest? Surely it cannot be a matter of hoping to construct a machine capable of generating well-formed literary "strings." What does any such formal modeling have to do with our *experience* of literature, with our subjective *understanding* of its multiple meanings and qualities? A literary theory does not try to substitute calculations for creation, deductions for readings. Nor does it want to replace criticism with a machine. Not only do we not want a ma-

ston (Saratoga, Calif.: Anma Libri, 1984), 177–89. A clear and useful presentation is that of Daniel Andler, "Les Sciences de la cognition," in *La Philosophie des sciences aujourd'hui*, ed. Jean Hamburger (Paris: Bordas, 1986), 131–75.

chine that will produce texts, but we do not want one that will read them. Why, then, would we want one that gives us the algorithm of the reader? But can the literary discipline specify its own standards of "rigor"? What are they? What is the literary definition of "theory"?

Theory as Epistemology

I have just set certain convictions about the nature of literature and its criticism in opposition to the very idea of seeking a formalized theory. These convictions hold that such a theory would be inappropriate—even should it be possible to devise one, which seems unlikely. Such objections bring to the surface a central feature of the debates over the role of theory within the literary disciplines, namely, the fact that what is at stake in many of these debates is the nature of the stance that the critic should take in regard to that vast and open-ended field known as literature. Moreover, what comes to the fore is the extent to which the literary critics' implicit and explicit decisions about the nature of literature and its study do not exist in a vacuum. What I mean to point out here is that the literary form of knowledge, a form of knowledge manifested in part by what critics do, exists alongside other types of knowledge in the academy and in society at large. Comparisons between these different models of knowledge are pertinent and necessary insofar as the different activities that they represent do not in fact exist in isolation from each other. On the contrary, they cohabit the same institutional spaces. That part of literary culture and of literary knowledge which exists within the modern university and which owes its very possibility to the existence of that institution cannot reasonably be held to be autonomous. Nor does it obey only its own self-assigned norms as to what constitutes a valid, worthwhile, or rigorous form of knowledge. Such points are elementary and should not have to be belabored. Yet some of the literary critics who identify something called "the institution" as the ultimate source of all critical authority and validity seem to be missing this basic fact, for in their discussions it appears that what is referred to as the critical institution only amounts to the public opinion of the professors of literature. Thus the larger institutional and social conditions that englobe the latter are left out of the picture. If everything is ultimately to be a matter of "the social," let us please have some serious sociology, and not another reworking of the literary and idealistic concept of "tradition," which refers to the relations between disembodied minds inside an ideal library.

The notion of "literary theory" that emerges in response to the question of the relations between literary and other models of knowledge is what I call an "epistemological" definition of theory. This book is meant to be a contribution to this type of work, and thus my understanding of the latter is reflected throughout these pages. One crucial feature of what is intended is an emphasis on the importance of reflecting on the nature and status of scholarly inquiries into literature. To engage in such metalevel reflection is not to place criticism in the foreground at the expense of literature, nor is it to efface the difference between the two; one cannot do an epistemological analysis of criticism without considering its relation to what it studies, which means that one must look not only at what the critics say but at the works they are talking about. But what is the point of this epistemological reflection? The briefest answer I can offer is that such a theory is antidogmatic. Without for a second wanting to assign theory a purely negative, critical, or destructive role (that of the perpetual ironist), I would like to stress first of all the importance of critical theory's challenge to the certitudes of substantive theories as well as to those of the invisible theories of "business-as-usual" literary criticism. One way a comparative literary epistemology, that is, one that looks at criticism's relationship to other models of knowledge, can effectively criticize the reifications of substantive theory is by contrasting the latter's claims to those of other discourses. Again and again a substantive literary theory makes claims concerning its own autonomy and supremacy which crumble in the face of any serious confrontation with research on related topics in other fields. It usually turns out as well that the substantive literary theory derived a good part of its material by means of a series of selective appropriations of some "theory" (generalities) circulating in other disciplines. Structuralism is a good example, for its proposed unification of the human sciences ultimately rests on a series of analogies based on some general ideas taken from a particular vein of linguistics—a vein of linguistics, moreover, that is highly problematic. In relation to invisible theory, then, the epistemological approach sets out to make the implicit and unconscious assumptions and habits visible and then pursues the question of their validity in as unflinching a manner as possible.

Such are two ways in which theory, construed as epistemological reflection, can intervene in literary research. It should be obvious that every good critic is engaged in the same kind of thinking on a regular basis, even if he or she never writes an essay on theory. As I have said, the contributions of this kind of approach are not purely negative, even if the more positive aspects are somewhat tenuous. In a sense this must be so, for the danger of collapsing into bogus substantive theses must always be kept in view. My own hypothesis is that, at the present stage, the most that a theory should attempt is a clarification of the topics and lines of inquiry available to critical research, a clarification that, if it rejects some types of work as ill founded and based on spurious goals and assumptions, actively encourages and contributes to other lines of inquiry. Herein resides a positive dimension of theory that is lacking in the ironic style, which tends to lend value only to its own negations of the pretensions of other texts and discourses.

I contend, then, that the most appropriate role of theory in contemporary literary study is an epistemological one. Theory should be concerned primarily with assessing the nature and status of literary knowledge—and should do this particularly in relation to the models of knowledge presented by other fields of research. Yet not all such comparisons are equally pertinent or equally pressing; we must recognize that the modern intellectual landscape is dominated by one model of knowledge—that of the natural sciences—and that this domination is so great as to put in question the very validity of other models of knowledge. A series of oppositions draws a dichotomy between the "hard" and "exact" sciences and their contraries, which are in the same stroke positioned as so many "soft" and "fuzzy" expressions or points of view. Thus any number of other domains and practices fall under the influence of an image of science which may in fact bear little relation to its reality. One of the central hypotheses of this book is indeed that much of critical theory is oriented by images of science that should not be taken for granted, and I shall be at pains to challenge the dogmatism of these views. Critical theory will thus be confronted with alternative conceptions of science, conceptions that no literary critic can dismiss with a casual gesture. Furthermore, by entertaining a different understanding of the natural sciences we may be able to arrive at some constructive clarifications of the nature of criticism's different tasks, including both those that it does and does not share with the general scientific project. Only in this way can an undistorted perspective on the values and specificities of literary knowledge be achieved.

At this point I think it necessary to lend a little support to my contention about the importance of a *comparative* literary theory.

One argument along these lines is quite simple: the comparison is already being drawn by critics on a regular basis, *et pour cause*, and thus I need only urge them to do a better job of it. A few examples are in order.

No one is surprised to discover that some of the romantics made wild claims in favor of poetry and art, which they typically opposed to the limitations of a mechanistic science. It would be wholly out of place to engage in a scientific evaluation of such statements—for surely the interest and value of the romantics' achievements lie elsewhere. When the writer is Shelley, it is neither astounding nor irksome to read such declarations as the following: "Poetry is indeed something divine. It is at once the centre and circumference of knowledge; it is that which comprehends all science, and that to which all science must be referred."[9] But it is rather different when we come across what looks like a gloss on these lines, asserted as a matter of truth, within one of the guidebooks published by the Modern Language Association. In reading the following, I can only note that the comparison between literary knowledge and other models of knowledge is being drawn, and that it is being drawn badly:

> Primarily under the aegis of comparative literature, the study of letters has become progressively more interdisciplinary as well as interliterary. Through this important development in modern scholarship, literature is being restored to its pristine position as a central cognitive resource in society, as its most faithful and comprehensive interpreter. It is an art but more than an art, for, while being itself, literature extends outside itself to forms of human experience beyond disciplinary boundaries, making it evident that the rigid separation of disciplines by myopic specializations can in the long run lead only to counterproductive and paralyzing isolation. Literature, as the hub of the wheel of knowledge, provides the logical locus for the integration of knowledge.[10]

These assertions do set the wheels spinning, and the only way I can bring them to a stop is to read the passage as a form of abnegation: neither literature nor its critical exegesis was ever central to the production of knowledge in the modern era, nor have they recently acquired any such status (the very real achievements of comparative literature notwithstanding). It is not the least bit obvious that litera-

[9]Percy Bysshe Shelley, "A Defense of Poetry," in *Critical Theory since Plato*, ed. Hazard Adams (New York: Harcourt Brace Jovanovich, 1971), 499–513, citation: 511.

[10]Jean-Pierre Barricelli and Joseph Gibaldi, eds., *Interrelations of Literature* (New York: Modern Language Association, 1982), iv.

ture—if by this we mean the kind of items studied by professors—is somehow central to our culture on the whole, or that it is in any way recognized as a "faithful and comprehensive interpreter." Yet we would like it to be, or at least we would like it to be somewhat less marginal and ignored than it currently is. But bald assertions do not make it so.

What, then, are more likely candidates for the role of hub of the wheel of knowledge? In the Western tradition, philosophy has been the academic discipline that has pretended to provide the logical and other loci wherein knowledge is integrated, and we need only look to the first pages of the *Metaphysics* of Aristotle to see an explicit articulation of the hierarchy of knowledges and practices.[11] At the top is a particular kind of *theoria*, which elevates itself above any of the special sciences in its grasp of first principles and ultimate explanatory factors (*archai* and *aitia*). It is philosophy that is supposed to provide the organon, classify the several senses of being, clarify the different types of wisdom, science, and technique, and even lay down the principles of poetics. Whether philosophy has ever really managed to do this is another question, and it appears that the special sciences in fact dethroned their queen quite some time ago, and that she now frequently seeks to play the role of handmaiden.[12] But in any case, we can hardly imagine that poetics has somehow taken her place. In this regard, we may grasp the inadequacy of the many debates in literary theory which square off only two terms—literature and philosophy—and which presume to deal in this manner with the fundamental issues concerning literary knowledge. As long as the sciences are excluded from the picture, the results can only be one-sided, for it is forgotten that philosophy, far from being the same thing as science, must deal with the problem of its own status in relation to scientific knowledge.

It hardly seems new or particularly insightful to point to Western culture's massive investment in natural scientific models of knowledge. It hardly seems a revelation to insist that in matters having to do with the validity and utility of claims to knowledge, it is a scientific model that is held up as the yardstick against which other

[11]Aristotle, *Metaphysics*, trans. Richard Hope (Ann Arbor: University of Michigan Press, 1960).

[12]In his excellent survey of the history of Germany philosophy between 1831 and 1933, Herbert Schnädelbach stresses the impact of the progress of the natural sciences upon philosophers' images of the status of their own discipline: "Philosophy was thus in the unfortunate position of having constantly to demonstrate its indispensability, or even its right to exist." See his *Philosophy in Germany, 1831–1933*, trans. Eric Matthews (Cambridge: Cambridge University Press, 1984), 67.

conceptions are measured. Nor has this kind of measuring been the work of only the natural scientists; it has also been regularly taken up by humanists—both those who celebrate the achievements of science, such as the Newton poets, and those who seek to oppose its murderous dissections. An example may help to illustrate what is meant here, for what I want to bring forth is the way the "defense of poetry" finds itself inscribed within a space where the scientific model has already been positioned as the source of illumination. Edgar Allan Poe, in what everyone instantly recognizes as a typically antiscientific and counter-Enlightenment gesture, wrote a "Mesmeric Revelation" in which a character under the influence of hypnosis has a metaphysical vision. In this vision, a solution to the matter–spirit dualism is glimpsed. Poe is proud of these ideas, which he casts in the form of a fantastic tale. He even toys with the possibility that they may be true and writes letters asking whether there may not be some confirmation of his notions.[13] As Poe's own antiscientific "Sonnet—To Science" suggests, the antimythical vulture's wings may be dull, but the poet nonetheless recognizes them as realities.

The literary and the scientific forms of knowing do not exist in isolation, which is not to affirm that they are not profoundly different or that either one of them integrates the other. Yet the profound influence of images of science upon the status of "poetic knowledge" must never be underestimated. Hence I would stress the importance of Stefan Morawski's observation that "the skepticism about esthetic inquiries started with the era of the paradigmatic domination of the exact sciences."[14] To speak of the paradigmatic domination of the sciences, then, seems perfectly apt, for it is the scientific model of knowledge which, even when misunderstood or wildly exaggerated, stands as the central image of what counts as knowledge's perfection. Nor is this only an abstract and general situation; any student of romanticism will affirm that biological concepts and images have played an important role within the literary culture of the past two hundred years, and we have the occasion to observe in subsequent chapters that these biological notions are hardly absent from recent critical thought.

[13]Thomas O. Mabbott, ed., *Collected Works of Edgar Allan Poe,* vol. 3: Tales & Sketches *1843–1849* (Cambridge, Mass.: Harvard University Press, 1978), 1024–42, esp. 1025n.

[14]Stefan Morawski, "What's Wrong with Aesthetics?" in *Die Ästhetik, das tägliche Leben und die Künste:* 8. *Internationale Kongress für Ästhetik* (Bonn: Bouvier, 1984), 29–37, citation: 29.

The *image* of science, then, stands as a crucial model of knowledge. What forms does the decisive image of science take? Two are singled out in the present context, for there are, I think, two particularly prevalent and disastrous ways of thinking about science which distort the humanists' views of the nature and status of their own research. Briefly put, the first takes us back to our definition of a formalized theory. Instead of imagining the difficult construction of such a calculus in relation to some restricted set of data, let us conceive of writing a formalized theory of "the universe," an algorithm capable of calculating all of the states of everything that is. Let us furthermore confuse this incoherent fantasy with science, and we may then arrive at the first image: the myth of a single, overarching and shadowless Science. That poetry has no place here is not astounding given that the illusory qualities of human experience have been reduced to a fundamental level of description, one where they appear as the mere effects of states of matter in motion. Yet the master theorist remains to manipulate the formulas, which enable him to predict and to dominate the states of nature. This Laplacean fantasy is obviously something the humanist wants to oppose, to refute, and to swerve away from in any way possible. And so one asserts the contrary: science cannot possibly be like that, science is shadowy too, there are only shadows, myths, visions, and uncontrollable dreams. Poetry cannot be measured in relation to the shadowless Science, because no such thing exists.

That science has its shadows remains an important observation. The natural sciences' massive complicity in the business of warfare, their essential and incessant contributions to what was once known as the military-industrial complex, their role in the technically efficient but uncontrolled exploitation of nature and of human beings—these are shadowy affairs indeed, and they are not merely to be discounted as accidental features of the presently instituted forms of scientific research. The idea that literary and, more generally, humanistic scholars should raise their voices in opposition to such realities seems crucial. Yet the specific forms this opposition often takes strike me as being wholly inappropriate and useless. To give only the broad strokes, one may characterize such shortcomings in the following way: having understood that, in relation to one view of science, literary and humanistic knowledge is nothing but a series of gratuitous worldviews, the literary scholar returns the insult, declaring that scientific knowledge too is nothing but a worldview. Here I arrive at the second prevalent image of science which plays an effective role in many literary debates, in spite of its erroneous characteristics—the idea that scientific knowledge is a myth. Such an opinion, which reappears frequently in a variety of shapes and sizes, typically rests upon a broader position, which I characterize as "framework relativism."

There may be other, more appropriate general names for this tendency—"historicism," "conventionalism," "constructivism," "idealist empiricism," or "cultural relativism"—just as there are various distinct subvarieties of this general stance, each having its own preferred vocabulary and emphases. The tendency I wish to identify, however, can be singled out by referring to a cluster of tenets that characterizes the diverse positions in question. Above all else, what I have in mind is a notion that reappears frequently in literary debates, namely, the idea that the "knowledges" produced by science have no truth other than that bestowed upon them by their own autonomous frameworks of concepts and prejudices. In its most tendentious form, this thesis amounts to the claim that the natural sciences have not achieved any real progress toward a true knowledge of nature; indeed, that the nature the sciences are wrongly thought to describe and explain is in fact determined—in some very strong sense of this term—by the changing presuppositions and languages ("paradigms") of the sciences. The theories and evidence of a given scientific explanation or description are said to be valid only in relation to such a framework, and as a result it is impossible to offer any rational basis for the idea that one explanation is more accurate than any other. Moreover, framework relativism contends that there exist radically divergent frameworks, so that beliefs that are true in one are false in another. It is consequently impossible to prove that natural science has progressed or is progressing toward truth, not only because the history of Western science reveals a discontinuous succession of frameworks or worldviews, but also because of the existence of totally nonscientific frameworks of belief. It follows that it is equally impossible to justify the notion that natural science has achieved more reliable truths than those claimed by any other cultural practice; accordingly, mankind does not know any more about nature now than it did four hundred years ago, before the emergence of modern science, and science is not at all founded in denying the validity and soundness of alternative conceptions of nature—those of so-called superstition and mythology. In other words, this camp's answer to the demarcation problem is that there can be no correct *epistemological* demarcation between science and nonscience. Freed from the illusion of a normative model of knowledge, literary critics can go on playing the game of interpretation, confident in the knowledge that their activity is one bundle of relativistic "language games" among others.

Is this merely a caricature of contemporary literary theory? I think not, for it can be shown that any number of prominent literary thinkers ascribe to this sort of position; more generally, it is simply a fact that they frequently relate their work to an image of science, and that when they do so it is more often than not some variation on the framework relativist scheme that stands in for the reality of natural scientific research. A demonstration that fully supports my assertion has recently been made by Paul Sporn, who shows not only that representatives of four major schools of critical thought have sought an orientation and legitimacy in certain interpretations of quantum mechanics, but also that the status of these analogies is by no means certain.[15] Additional examples can be easily provided; I limit myself in this context to a few, chosen in function of what I perceive as some of the principal features of framework relativism. My examples may usefully be read against the background of what now represents the *vieux jeu*, the first page of Lucien Goldmann's essay on Racine, which begins with the observation that there is a lag between the results of the exact and the human sciences; moreover, within the latter, that there is an analogous lag between sociology or history and the study of literature.[16]

My first example is an article that expresses a "deconstructive" perspective on the nature of literary critical knowledge. This seems to be a perspective concerned with the danger that critics entertain false beliefs in the universal and literal truth of their interpretive statements.[17] To escape the error of taking its own constitutive metaphors literally, literary criticism must position itself within an "Einsteinian" space. What does this mean? The author asserts that, since Einstein, we have learned that the choice of a frame of reference, a choice from which everything else is said to follow, is and can only be arbitrary. For this reason it is necessary to recognize the fundamental relativity of all readings and interpretations. But such a recognition appears to have paradoxical consequences, for this recognition itself must only be a metaphor. Thus the essay concludes by turning upon itself to assess the status of its own assertions: "Like

[15]Paul Sporn, "Physique moderne et critique contemporaine," *Poétique*, 67 (1986), 315–34.

[16]Lucien Goldmann, *Racine* (Paris: L'Arche, 1956), 13.

[17]I do not discuss whether the article in question is truly exemplary of deconstruction. See Gasché, "Deconstruction as Criticism," as well as my remarks on Spivak, Nancy, and Derrida in what follows.

the terms 'literal' and 'metaphorical' crossed out above, this essay appears only to disappear again. And the cost of its reappearance seems to be nothing less than this disappearance."[18]

My second example illustrates another typical point, one often thought to follow upon the first. If all frames of reference are arbitrary, individuals must have the freedom to move from one perspective to another. But then it hardly seems that any of these frameworks can be binding, or that one's worldview or reality can be determined by a framework that can easily be replaced by another one. Here the typical response is that the frameworks in question are institutional, and thus that the institution is responsible for all validity. The individual's beliefs are thus determined by the institutional framework within which intellectual practices are alone possible. As institutions vary, so do norms of validity. Here is an example: "The assumption that methods common to scientific inquiry are also objectively valid methods for criticism is wrong because the standards and conventions of scientific inquiry have their validity within the institution of science."[19] This is anything but a ridiculous assertion, and as I point out at some length in the next chapter, some well-informed students of science take such a notion quite seriously. But it is not an established truth, either, and no serious discussion about the possibilities for literary knowledge can simply take it for granted. There are some real problems in the purely sociological accounts of science's validity, and philosophers of science have mounted strong arguments against the institutional hypothesis.

Literary critics often simply take it for granted that it is correct to conclude that the determining framework of science is an arbitrary institutional fact. Yet even this is not enough to dislodge the mythical certainties of natural science, and thus literary critics often go on to claim that science, far from possessing any unity or progress, is really a matter of a series of discontinuous and heterogeneous frameworks. The only validity that science can have—which ultimately is no validity worth having from a scientific perspective—is that granted within one of a variety of paradigms. Moreover, these paradigms are so divergent that what figures as true science within the one is falsehood and mythology within the next. Here the literary theorist calls upon those expert witnesses, Thomas Kuhn and Paul Feyerabend, whose writings can be read as lending the needed sup-

[18]Michael McCanles, "Criticism Is the (Dis)closure of Meaning," in *What Is Criticism?* ed. Paul Hernadi (Bloomington: Indiana University Press, 1981), 268–79, citation: 276.

[19]Suresh Raval, *Metacriticism* (Athens: University of Georgia Press, 1981), 159.

port. Thus we arrive at what appears to be the unassailable *doxa* of framework relativism. My next citation is from an intelligent attempt, on the part of a literary scholar, to think through the most extreme implications of these tenets: "Am I suggesting that, if it is impossible to prove or disprove a paradigm, and if communication is impossible between proponents of different paradigms, that 1) traditional and post-structuralist critical theories are equally valid? and that 2) theirs is a hopeless argument? In a certain sense, yes. For, according to the post-structuralist paradigm at least, no theory can be *absolutely* valid, that is, *True*; and it is certain that no paradigm can be 'understood' from within the conceptual domain of another."[20]

In the typical appropriations of Kuhn and Feyerabend, what is asserted is the irrationality of science, its historical discontinuities, its lack of a grounding in some veridical reference to an invariant nature existing outside of its frameworks. Bound within one such framework or paradigm, the scientist does not have the freedom to adjust his or her thoughts, reflections, and beliefs to reality, and thus it is pure illusion to hold to the viability of a project whereby the conceptions of scientists are progressively being matched up with the features of an objective nature.[21]

In some of the more "philosophical" versions of this kind of stance, the thesis of the *Sein-und Zeitsgebundenheit des Geistes* rests upon some rather far-reaching assertions about the course of Western history. It is important to realize the extent to which many thinkers' convictions about framework relativism are supported by such assumptions. An exemplary statement of these views is that of Jean-Luc Nancy's *L'Oubli de la philosophie*, a work that is presented by its author as voicing, in a didactic spirit, what should be known and recognized by anyone in touch with contemporary philosophy.[22] And thus with an admirable frankness and unflinching confidence, Nancy sets forth the theses of a perspective that finds numerous

[20]Candace Lang, "Aberrance in Criticism?" *Substance*, 12 (1983), 3–16, citation: 14.

[21]For an intelligent discussion of references to the concept of "paradigm" in German literary-critical circles, see Kurt Bayertz, *Wissenschaftstheorie und Paradigmabegriff* (Stuttgart: Metzler, 1981), 106–18. The author specifies conditions for an appropriate usage of this term and gives reasons why these conditions are not satisfied in the literary examples he cites.

[22]Jean-Luc Nancy, *L'Oubli de la philosophie* (Paris: Galilée, 1986). This essay is presented as a rejoinder to various unnamed critics of an unspecified group of French philosophers, the principal object of the attack being undoubtably Luc Ferry and Alain Renaut, *La Pensée 68: Essai sur l'anti-humanisme contemporain* (Paris: Gallimard, 1985). My criticisms of Nancy should not be taken as the sign of any agreement with Ferry and Renaut.

manifestations in literary theoretical writings. To begin, there is the assertion that, if it once made sense to believe in the progress of knowledge toward truth— or in the progress of civilization toward its ideals—any such credence is today an imposture and amounts to a "forgetting of philosophy." By this is meant the forgetting, not of philosophy's initial projects and aspirations, but of what has become of them. For since Kant's time, "something has happened," and this something can be signaled by such names as Wittgenstein, Heidegger, Mallarmé, and Einstein.[23] The Enlightenment project has played itself out; we have seen the end of metaphysics and the death of God. With these events has died the project of a humanism whereby mankind would finally become "himself" as the full embodiment of Reason. Once again, science is never really discussed, yet the word is twice included in the list of discredited ideals.[24] It is here that an important objection must be made. One may grant that it is an open question whether the history of the past two hundred years warrants any belief in mankind's progress toward goals of liberty, freedom, and equality. The gruesome realities of warfare, tyranny, and barbarism encourage at the very least a prudent agnosticism on such an issue (nor is it clear following what inductive principles it has been decided that today determines the historical point from which accurate generalizations concerning history's "necessities" can be made). But it is another matter entirely to extend this analysis immediately and *sans phrase* to the case of scientific knowledge, which is what Nancy does. Science may not have realized its ideal, nor has it necessarily made humanity happier or freer, but it is sheer dogmatism simply to assert, on the basis of references to Kant, Heidegger, and Wittgenstein, that no significant scientific discoveries have been made over the past four hundred years. In fact, Nancy's defense of such a judgment concerning the fate of the scientific project is supported only by his broad analysis of the type of "desire" that he feels is responsible for the organization of all such "projects."[25] The desire in question is that of a divided subject, a subject containing within itself a fundamental lack, for its truths, value, and identity are posited as being outside it and reachable only at the end of an infinite striving. The scientific fiction of an asymptotic progress toward an unrealizable ideal of perfect exactitude and a global, deterministic knowledge, then, reflects this same structure of desire, with its basic and insur-

[23]Nancy, *L'Oubli de la philosophie*, 25, 68, 80.
[24]Nancy, *L'Oubli de la philosophie*, 46, 72.
[25]Nancy, *L'Oubli de la philosophie*, 43–52.

mountable duality of the "lack" and the "project." As such, the specific details concerning any of science's achievements or discoveries do not address the essential, for they cannot possibly embody the ideal. One simply cannot get any closer to an infinitely remote ideal. Yet it is possible to realize the futility of the striving, to suspend and interrogate the basic structure of the desire which gives rise to it—which is what such figures as Wittgenstein and Walter Benjamin are said to have done.

Even if we grant that the status of natural scientific knowledge can be adequately analyzed in terms of the nature of the "desire" that animates scientific research (a step requiring no small assumptions), there remains a basic error in Nancy's discussion. He examines the ideal of science, a model of perfection infinitely removed from the particular states of scientific knowledge, and he then makes the observation that the actual states *are not* the ideal. The flaw here is that this analysis obscures the existence of other ways of relating the actual to the ideal, ways not exhausted by what Nancy would call the "copula." Suppose that k1 stands for the state of medical science before Harvey, and let k2 stand for medical science after his contributions; finally, let K stand for the ideal of a perfect and total knowledge of the workings of the human organism. Enter Zeno, who observes that k1 is not K, that k2 is not K, that both are infinitely removed from K, and that therefore we may conclude that it is nonsense to say that k2 is further along toward K than is k1, or that reaching k2 amounts to progress in medical knowledge. The error here is that of thinking that the relation between the individual k's can be evaluated only in relation to the possibility of fully actualizing the ideal, or K. In fact, the value of the project, and with it the meaning of the notion of progress, resides not in the ideal as such but in the approximation of that goal, and it is in regard to this approximation that k2 amounts to real progress over k1, for k2 is in fact a better approximation of the truth than was k1. Harvey's discoveries did not, to be sure, amount to a realization of the perfect knowledge of the human body, but they were discoveries, and to obscure this general category of historical fact out of a preference for ontology is very bad philosophy indeed. If science must be said to "desire," what it desires is the best possible approximation of an explanation that, although forever beyond reach, can provide an effective guideline for the progress of human knowledge.

To sum up, I have tried to show that the cluster of positions I characterized as framework relativism is not a figment of my own imagination but figures as a major tendency within literary theoretical circles.[26] In the following chapters, I argue at some length and in detail against this tendency, for I am convinced that its dogmatism can only be harmful to the possibility of elaborating and defending literary knowledge.

But who said knowledge was so important, anyway? In response to this question, which one hears often enough in literary circles, it is appropriate to discuss some other options for literary theory.

Theory as Art, Theory as Fun

Everything I have said so far takes for granted the hoary assumption that literary theory should interest itself in arriving at knowledge. To put this another way, my epistemological reflections take for granted the idea that critical work should be viewed primarily as inquiry or as research. This assumption is made throughout this book. But those who are less sanguine about the defense of literary knowledge, or about knowledge in general, exhort us to make a basic shift in perspective. Critics should stop being so concerned with knowledge, they say, particularly the sort that orients itself toward the fantasy of truth. Are there not other games we can play? Indeed there are, and the different games being proposed all set aside the traditional goal of scholarly work, the "original contribution to knowledge," which is judged to be either impossible or simply undesirable.[27]

I see nothing wrong in principle with the idea that reading and writing literary criticism should be creative and pleasurable experi-

[26]So much so that one looks far and wide for exceptions. One major exception, however, is that of Siegfried J. Schmidt, who has made impressive contributions to a literary theory that remains in touch with developments in the philosophy of science, cognitive psychology, systems theory, and sociology. Although I do not ascribe to all of his views, may the value of his work be acknowledged here. See his *Elemente einer Textpoetik: Theorie und Anwendung* (Munich: Bayerischer Schulbuch, 1974); *Literaturwissenschaft als argumentierende Wissenschaft: Zur Grundlegung einer rationalen Literaturwissenschaft* (Munich: Wilhelm Fink, 1975); *Grundriss der empirischen Literaturwissenschaft*, 2 vols. (Braunschweig: Vieweg, 1980–82); "Selbstorganisation-Wirklichkeit-Verantwortung: Der wissenschaftliche Konstruktivismus als Erkenntnistheorie und Lebensentwurf," *LUMIS-Schriften* 9 (Siegen: Lumis Institut für Empirische Literatur- und Medienforschung, 1986). In English, one may consult "Literary Science as a Science of Argument," *New Literary History*, 7 (1975), 467–81, as well as various contributions to *Poetics*, of which Schmidt is the editor.

[27]I am not going to try to list the many texts that fall beneath this general rubric. Roland Barthes's *Le Plaisir du texte* (Paris: Seuil, 1973) was, of course, pivotal in its rejection of the goal of explanation in favor of hedonistic and personal appraisals of the different pleasures and *jouissance* to be had from reading literature.

ences. In fact, both of these values are crucial. Yet I do not think that pleasure and art are priorities that should replace the goal of inquiry and the epistemological norms that follow upon that goal, nor do I think that they should be allowed to vitiate the attempt to move toward such goals. Perhaps in some truly utopian society the distinctions between work and play, analysis and fantasy, arguments and images, and so on, will be dissolved, and in that world the question will have vanished. But we do not live in such a world. Until we do, there is a better way of replacing the hoary notion of "knowledge for its own sake."

Theory as Social Action

The tasks of criticism should, in my view, be differentiated from each other as carefully as possible, for progress along the different avenues of research requires a clear sense of orientation. As I argue in my final two chapters, criticism often falters because of a fundamental confusion about what tasks are being aimed at. Moreover, critical inquiry could be oriented toward very different sorts of goals. All forms of inquiry are necessarily selective and seek to attain particular ends. All forms of knowledge share a first-level, and very general, value orientation in their common goal of arriving at the best possible approximation of the truth. But this first orientation is neither restrictive nor determinant enough to establish a particular set of priorities, even less a list of topics or questions. The possible range of meaningful and coherent lines of inquiry remains vast, indeed, infinite. To contend, then, that literary studies should find an additional orientation in the needs and problems of the society within which it exists is in my view an important stance. This stance responds to the need for a second-level value orientation guiding all research. But of course other stances could be substituted for it. One could stipulate that inquiry be guided first by its search for the truth and, second, by the individual scientist's "rational" goal of maximizing personal utilities (the transcendental signifier of which, in our world, is the salary). One could also argue that this latter position amounts to the best way of construing the first stance, for will not our "unsociable sociability" produce the maximal social justice in the long run? I, for one, do not think so, and I believe that there are better ways of trying to understand the first stance. I do not think that personal ambition, habit, and blind tradition should be left to guide decisions about the course of literary research, decisions concerning its choice of questions, priorities, and avenues of inquiry. Far too many publications appear in which the importance that is supposed to be attached to the result is a great mystery; critics would do well to make the nature of their various claims and ambitions clearer.

What I have just said, however, should not be confused with any narrow instrumental attitude toward knowledge, for one of the pressing needs of our modern societies is to deny the imperatives of crassly pragmatic and productivist ideologies and the appropriative and exploitative institutions they support. In a world where business and the military frequently stand as the greatest national priorities, literary criticism's pursuit of "useless" truths, its insistence upon remembering the past for its own sake, are virtues. Yet in spite of this recognition, my deeper conviction is that literary knowledge is best when it actively and directly articulates its opposition to the present world order, that is, when it attempts to comprehend and to remedy what is wrong.[28]

[28]I do not in the present work expand on my opinions on this subject, although it becomes clear what I think is wrong with literary research, and what alternatives should be envisioned. To lapse momentarily into the prophetic and substantive mode of theorizing, I predict that when future critics look back upon the present period of work in the literary disciplines, the most important contribution to be noted will not be that of the skeptics and their textual gymnastics; rather, what will clearly stand forth as a necessary and constructive shift in the literary fields will be the emergence and institutionalization of research, pedagogy, and other practices oriented by feminist perspectives.

PART TWO

OF SCIENCE AND NATURE

In my discussion of some of the basic conceptions of knowledge entertained by literary theorists, I isolated one position characterized as "framework relativism." I have suggested that, if framework relativism is by no means a dominant voice within the institutions of learning and research, it nonetheless remains an extremely influential tendency, particularly within those circles that seek to engage in an extended theoretical and interdisciplinary reflection upon the nature and status of literary inquiry. In a sense, the arguments of the framework relativist are of great value, for they raise issues that badly need to be confronted by the literary disciplines, which have long suffered from a form of naive empiricism. At the same time, however, the framework relativist doctrines are sometimes asserted in a dogmatic, superficial, and seemingly automatic manner that only serves to obscure the alternative lines of inquiry available to literary critics. As I have suggested in the last chapter, this kind of dogmatism is particularly accentuated in regard to the key issue of the literary critic's understanding of the natural sciences and of the methodological models associated with them. For this reason it seems crucial in the present conjuncture to reexamine these issues in greater detail.

What is at stake, then, in the next two chapters is the general account of knowledge and of natural science that should figure in an understanding of literary knowledge. Consequently, I explore some central topics in the theory of knowledge and in the philosophy of natural science. One of the most important questions to be considered is whether the positions associated with framework relativism in fact offer the best available account of the natural sciences, as literary critics frequently assume. My own belief on that score is that they do not, and so I compare the framework relativist's perspective on natural science to other positions, and in particular, to a realist account that distinguishes between science and nonscience on the basis of *epistemological* differences. In other words, I explore arguments by means of which natural science could be distinguished from other symbolic forms by virtue of the nature of its knowledge claims. Although it is clearly impossible for me in the present context to offer any definitive and comprehensive description and defense of realism in regard to the natural sciences, it is my goal to highlight the strengths of this general type of approach to science. I especially want to challenge the adequacy of certain literary versions of the philosophy of science, versions in which a vast and complex field is reduced to some of the more tendentious conclusions that can be pulled from the writings of Kuhn and Feyerabend. Literary critics should be aware that this manner of relating their own theoretical reflections to science really amounts to a highly selective reference to the philosophy of science, at a time when more recent developments in the latter field hardly support such a practice. Although I am far from asserting here that any of the realist accounts are definitely correct, I do want to insist that literary theorists must look beyond the sterile opposition between positivism and framework relativism if they want to know something about the philosophy of science—and beyond it, about the sciences themselves.

What, then, is this vein of recent philosophy of science that has been overlooked? Briefly put, it is a matter of accounts in which the tangible and prevalent success of science in our world is partly understood in terms of the scientific method's reliability in producing realistic explanations of nature. This means, for example, that the nature referred to in natural scientific explanations is a reality independent of mind and culture; furthermore, the nature in question is taken to be a reality knowable by human beings, albeit always partially and by successive approximations. Scientific explanations are successful, then, insofar as they refer—partially and with approximate truth—to a mind-independent nature not entirely caused, constituted, constructed, or overdetermined by the framework of science itself (be this framework understood as that of mind, theory, language, paradigm). These views, then, amount to a refutation of the humanistic and anthropocentric view whereby nature is entirely a human artifact, the product of mankind's own shifting frameworks—as well as the view that nature is an unknowable Being secretly at work around us yet forever hidden by some remarkable veil (a veil that is all the more remarkable because we are somehow aware of its presence).

At the same time, however, the realist position to be defended here must be distinguished in many important ways from both the classical and positivistic accounts of natural science which, in my view, are even less adequate than the "worldview" account under criticism. First of all, defense of a realist philosophy of science does not entail defense of reductionist materialism, or of the view that the goal of science is to discover the laws of a single, fundamental or ultimate level of description. Nor does the realist account to be discussed here assume that the philosophy of science can or should provide an algorithmic model of explanation. Moreover, a realist philosophy of science need not be committed to the erroneous idea that science is *only* a pure, epistemic activity, animated solely by a disinterested quest for truth, nor does it have any necessary link to the idea that scientific progress *equals* progress in any broader ethical or political sense. This point should perhaps be underscored because it is frequently overlooked by romanticist opponents of "Enlightenment myths" of rationality and progress: the question of whether a given scientific doctrine is true or refers (albeit approximately) to some natural reality is distinct from the question of whether acquiring such a doctrine is a good thing, not because science is value-free but because knowledge is one value that should not be equated with other kinds of values. This crucial point is discussed in Chapters 3 and 5.

Before the different accounts of natural science are taken up directly, a much more basic and general difficulty must be acknowledged. This difficulty concerns the criteria to be used in evaluating the understandings of natural science that are available today. It is important to address the status of the metalanguage that must be employed, explicitly or implicitly, in any deliberation over the relative validity of different interpretations of natural scientific knowledge. Do not all claims about scientic knowledge necessarily involve a number of assumptions about the world as well as norms specifying what counts as real evidence and as valid argumentation? Yet what are such norms and assumptions based on? Are they "scientific?" These questions raise the problem of whether it is possible to engage in a discussion of knowledge that does not already presuppose its own successful outcome. Does such a discussion rely in advance on the very model of knowledge it purports to discover? This is a perennial problem of epistemology; it is the problem that haunts any project of a transcendental philosophy that sets out to establish a priori limits and foundations for knowledge having a necessity and universality "prior to" particular attempts to know.[1] The difficulty would seem to be aggravated in the present context by the variety and number of the discourses that already claim to offer the most valid approach to science. The literary critic who would like to understand the model of knowledge represented by the natural sciences cannot turn to some unified and systematic philosophy of science; indeed, he or she is likely to have the impression that all that will be found there is the same Babel that characterizes the literary disciplines.[2] Moreover, in addition to a divisive literature on the philosophy of science, there exist any number of rival accounts of science arising from within other well-established disciplines, in particular, sociology and history. Nor can the problem be simply avoided by turning away from these metalanguages toward the scientists themselves, not only because the scientists also present very different accounts of their own knowledge, but because such accounts themselves are a matter of so many philosophical, sociological, psychological, and historical attempts to interpret science.

These are serious and far-reaching problems, and it would be simplistic to imagine that an answer to them is readily available. As many of the present discourses about the scientific model of knowledge simply ignore these much more basic problems—problems that put in question all of their supposed results—it seems crucial to deal with them first and as directly as possible. For this reason I devote some time to a discussion of basic epistemological issues before moving on in Chapter 3 to arguments about the natural sciences. I must once again address the problems related to the status of an epistemological metalanguage when I turn to the issue of the relation between the natural and human sciences in Part Three. First, however, I must discuss some skeptical arguments that seem to pose serious difficulties for the very idea that it is possible to have justified

[1]Robert B. Pippin, *Kant's Theory of Form: An Essay on the Critique of Pure Reason* (New Haven: Yale University Press, 1982), 222–23.

[2]This motif has been taken up recently by Michel Serres in "Dream," in *Disorder and Order: Proceedings of the Stanford International Symposium, Stanford Literature Studies*, 1, ed. Paisley Livingston (Saratoga, Calif.: Anma Libri, 1984), 225–39. A more Borgesian conclusion is drawn by the authors of an introductory work on the theory of knowledge: "Progress in philosophy has rather resembled the exploration of some enormous labyrinth in which we are all imprisoned and whose overall plan we do not know"; D. J. O'Connor and Brian Carr, *Introduction to the Theory of Knowledge* (Minneapolis: University of Minnesota Press, 1982), 188.

and well-founded beliefs at all. This discussion of skepticism is not undertaken for its own sake; it can help us to come to an understanding of what is at stake in the idea of justified belief or knowledge and also can provide a broader context for an understanding of today's prevalent framework relativism. A distinction between two basic kinds of skepticism emerges from the discussion, and it is possible to show that, although there is a coherent form of skepticism, framework relativism is an incoherent form of skepticism. After presenting this argument, I turn to the other obstacle to the idea that critical theory can play an epistemological role, namely, the evacuation of epistemology's very raison d'être in the so-called sociological turn.

CHAPTER 2

In Defense of Epistemology

It is fashionable today to announce the uselessness and failure of epistemology as such. The enterprise for which this word stands, it is held, has revealed itself to be as hopeless as it was empty. The ensuing statements concerning the consequences of this "discovery" of the vacuity of epistemological reflection may vary, but the opening gambit does not: the dream is over, the epistemological game is finished; what we should realize now is that the real initiative has been passed along to other disciplines and other preoccupations.

Yet what is such talk all about? Is epistemology truly a piece of equipment that a scholar or scientist can choose to take along or leave at home, following the circumstances? Is it really only an encumbrance—the umbrella incongruously brought along on the sunny day of literary criticism? It is true that some who have been announcing the end of epistemology and the uselessness of theory would seem to believe that the critical establishment should expect unending sunny weather. But the questions traditionally associated with epistemology will surely outlive such forecasts.

The Problem of Knowledge

What are these questions that epistemology is about, questions that are not imposed upon the hapless critic as if from without, but which emerge from within the practice of all inquiry? Let us begin to see how epistemological problems work by taking up some of the implications of the argument that was touched upon above, namely, the idea that the massive disagreements in the literature on science imply the absence of any true knowledge as to its nature or status. Sometimes the fact of the disagreements alone is taken as definitive proof that it is hopeless to imagine that anything has ever been established in this domain—or could ever be established. What, however, is the implicit definition of "knowledge" that underwrites such an attitude? Some sort of consensus theory of truth appears to be at work here, such as the idea that what is true is what everyone agrees upon ("mankind as a whole is the measure of all things"). Yet how could we know that such a model of knowledge is the right one—that it is true? Presumably it would have to satisfy its own criterion, so that it would be right if and only if everyone agreed to it. But how could we know whether it is true that everyone agrees to a consensus theory of truth? Here we see that this little caricatural "theory of knowledge" cannot ground itself, for it needs at least one fact about the world to get started, or to provide a basis—namely, a fact about what everyone believes. Yet its own definition of what constitutes a truth seems to prohibit it from obtaining such a fact. What is more, if we admit the truth of what seems like a pretty plausible fact—that the author of the present book does not agree with a consensus theory of truth (and I do not)—then we see that such a theory is flatly contradicted on its own terms.

One way to try to save this kind of theory is to introduce limitations about whose agreement is necessary. Perhaps it is too much to ask that "everyone" agree; besides, it is unclear whom that vague notion includes (must we count in the opinion of Attila the Hun?). Perhaps the consensus should be only that of the "experts," or "the ideal community of observers in the long run," or "people who obey the norms of reasonable dialogue." Yet now the problem of justification or of foundations returns to haunt us, for we must give a basis or justification for the criteria used to establish the limits to the consensus. To justify such limitations, one must have recourse to criteria that have nothing to do with consensus as such, so that it is not the consensus that is the ultimate basis of knowledge but such notions as "reason," "expertise," and so on which lend their deeper authority to whatever consensus is established. The results of the "ideal speech situation" are not valid only because they are the results of a dialogue or an agreement, but because the process is judged to be ideal in function of norms that are not themselves arrived at in the dialogue. In this model of knowledge, it is these norms that justify and ground the dialogue, not the other way around; the norms are there from the start, for they are held to be implicit in the true nature of language and undistorted communication. Although they can be disobeyed, the ultimate validity of these norms cannot be challenged, for they are supposed to provide the ultimate yardstick in relation to which such violations are measured. This is why Karl-Otto Apel refers to these norms as the "transcendental a priori" of communication.[1] Another argument popular today limits the consensus model not by seeking a truly valid or correct norm but by pointing pragmatically to the ways validity is established by something called "the institution." For example, what is valid in literary debates is said to be determined entirely by the consensus of those who shape the dominant and (and perpetually sunny) institutions where these debates take place. But what validates the institutions? Here it is recommended that the reader reflect upon Cornelius Castoriadis's insightful definition of "institution": an inertia supported by a myth.

These basic epistemological difficulties have been analyzed by a variety of thinkers who have sought to identify the logical structure of the problem.[2] In the tradition that begins with Plato's *Theatetus*, philosophers attempt to state the necessary and sufficient conditions for its being true that someone knows a given proposition; in other words, it is a matter of trying to formulate a logically rigorous, minimal definition of a bit of knowledge. Very basically, the classical account of knowledge that seeks to respond to this demand specifies that it is not enough for someone to be right (or to have a true or accurate belief) for us to dignify this as real knowledge; not only must the subject know, but the subject must know that he or she knows. Typically, this amounts to the problem of *justifying* the belief. Thus, in what is referred to as the standard account of know-

[1]See primarily the articles in the second volume of *Transformation der Philosophie: Das Apriori der Kommunikationsgemeinschaft* (Frankfurt: Suhrkamp, 1973); in English, *Towards a Transformation of Philosophy* (London: Routledge & Kegan Paul, 1980).

[2]I am especially indebted to the following sources: Rosalind S. Simson, "An Internalist View of the Epistemic Regress Problem," *Philosophy and Phenomenological Research*, 47 (1986), 179–208; Robert K. Shope, *The Analysis of Knowing: A Decade of Research* (Princeton, N.J.: Princeton University Press, 1983); Robert Nozick, *Philosophical Explanations* (Cambridge, Mass.: Harvard University Press, 1981), 167–290; A. C. Grayling, *The Refutation of Skepticism* (London: Duckworth, 1985); Ernest Sosa, "The Foundations of Foundationalism," *Nous*, 14 (1980), 547–64; W. W. Bartley III, "Non-Justificationism: Popper versus Wittgenstein," in *Epistemology and Philosophy of Science: Proceedings of the 7th International Wittgenstein Symposium*, ed. Paul Weingartner and Johannes Czermak (Vienna: Hölder-Pichler-Tempsky, 1983), 255–61.

ing, knowledge is analyzed as "true and justified belief," and an adequate specification of its conditions would constitute, in addition to a theory of truth (for example, a detailed account of the nature of true belief), a theory of justification. But it is precisely the possibility of such a justification which many people have challenged.

According to Hans Albert, every attempt to fulfill the claims for a fundamental grounding (in the sense of a *principium rationis sufficientis*) leads to "Münchausen's trilemma", a choice between the following three unsuccessful options:

1. an *infinite regress* that appears to be demanded by the necessity of going always further back in the search for reasons, but which is not practically feasible and therefore yields no solid foundation;
2. a *logical circle* in the deduction which results from the fact that one is forced in the grounding process to resort to statements that have already shown themselves to be in need of grounding—a process that, because it is logically faulty, likewise leads to no solid foundation;
3. a *cessation of the process* at a particular point, which is in principle feasible but which involves an arbitrary suspension of the principle of sufficient grounding.[3]

Albert sets forth this trilemma as the dire alternative to his own Popper-inspired position, which is that of a "non-foundational" or "pancriticist" attitude whereby no aspect of our knowledge is deemed to be basic and unassailable and where everything is open to criticism. The dogged traditional epistemologist of course wants to know what presuppositions and logical assumptions are taken for granted in this conception, but what interests us are the implications of Albert's trilemma for our basic presentation of the problem of justifying knowledge. Albert's schema indeed helps to delineate a major problem, yet it should be possible to focus the issue even more sharply. It seems that his three options can be reduced to two, for points (1) and (3) both involve the more basic idea that the chain of justifications cannot be left dangling, which is what results when it is broken off at some arbitrary point. Moreover, one might add that even point (2) amounts to the same problem—for here is what the circularity amounts to: on the one hand, we say that a proposition *p* grounds a proposition *q* but itself lacks a ground; on the other hand, we say that *p* is supposed to ground *q*, while the only ground for *q* is *p*. But the common assumption is that every proposition needs a

[3]Hans Albert, *Traktat über kritische Vernunft* (Tübingen: J.C.B. Mohr, 1968), 13.

ground or support outside itself in order to be justified and that a proposition is not really justified if its ground is a proposition that is not grounded. So what we really have to deal with is ultimately not a trilemma, but one epistemic difficulty: how can propositions be self-grounding? how can they provide their own confirmation or verification? In the theories of justification which are known as "foundationalist," it is asserted that the members of a particular class of beliefs indeed have this special status and consequently require no further justification. This class of beliefs is typically identified by appeals to states of "immediate apprehension," "direct awareness," or "self-evidence"; for example, the so-called internalist views identify a subset of one's own mental states as being indubitable, and thus as serving to justify other beliefs. Yet the critics of foundationalism want to know how this classification of the foundational beliefs is itself to be justified. Is this classification itself one of the basic, indubitable beliefs? The typical candidates proposed as basic, self-justifying beliefs within the empiricist tradition are notions associated with direct sensory or perceptual experience, and it is difficult to see how these experiences can carry within them in an immediate manner the conceptual and classificatory information necessary to their justificational role. How do I know, when I am having a basic and indubitable perceptual experience, that this experience is a member of the special class of foundational beliefs?

Rana Paradoxa, or the Skeptical Challenge

At the risk of belaboring the obvious, I would like to try to give the reader some sense of why these matters are important by means of a simple example (an example to be put to other uses as well). Many commercial developers have thought, and still claim today, that the natural environment can in no significant way be permanently damaged by the deforestation of parts of the vast expanses of tropical forests (for example, in Brazil). Yet some naturalists are of the opinion that a very large number of unique species inhabiting these regions will be permanently lost as a result of such activities. Given that present estimates indicate that roughly 100,000 square kilometers of tropical forests are being felled each year, it may eventually be possible to present approximate calculations of the number of species being eliminated. At the same time, naturalists cannot presently tell us how many different species exist on our planet; the estimates vary from three million to ten million, with roughly one and a half million presently classified).[4] Let us suppose that some naturalists have to justify their claim that a certain species of frog is endangered, while those who want to continue deforestation present counterarguments. It would be naive to think that people engage in destructive commercial activities only out of ignorance of their consequences, just as it would be foolish to assume that "the best argument alone" will decide what happens. Yet it seems a terrible exaggeration to think that argument has nothing to do with such a social issue. Certain factual beliefs are used both to support and to criticize the deforestation—beliefs about what its consequences really are, for example. Any serious argument over those beliefs must take into account the issue of their validity, that is, the question of how we could be justified in claiming to know which beliefs are true. In our example, it matters whether the creatures referred to in the ecologist's arguments exist or not; we can easily imagine someone justifying the current practices on the grounds that the ecologists have only invented a myth of endangered frogs out of some romantic attachment to tropical forests; the ecologists in turn would try to provide arguments in which they set forth reasons and evidence supporting their case.

The notion of evidence seems crucial in regard to the whole problem of knowledge, and it is important to resist any temptation to conceive of it in any unexamined manner, such as the idea that only an immediate and direct perception counts as evidence. Can we perceive or observe directly a species' endangeredness? Clearly not. The consequences are double. On the one hand, we must note the weaknesses of any naive forms of foundationalism in which a special class of basic perceptions or direct observations are set forward as an autonomous ground of evidence in all epistemological problems. Yet on the other hand, there are more positive consequences. The example suggests that it is crucial to hold open the possibility that certain types or forms of reasoning or inference also serve as evidence in inquiry. Most generally, evidence is to be understood as a belief or set

[4]For background on deforestation and endangered species, see Erik Eckholm, *Disappearing Species: The Social Challenge* (Washington: Worldwatch Institute, 1978); and Norman Myers, *The Sinking Ark: A New Look at the Problem of Endangered Species* (Oxford: Pergamon, 1979). On the highly problematic question of the number of species, see George Innis, "Numbers of Species and Optimization in Biology," in *Structure, Function, and Management of Ecosystems: Proceedings of the First International Congress of Ecology*, ed. A. J. Cavé (Wageningen: Centre for Agricultural Publishing and Documentation, 1974), 384–87; and the classic article by G. E. Hutchinson, "Homage to Santa Rosalia, or Why Are There So Many Kinds of Animals?" *American Naturalist*, 93 (1959), 145–59.

of propositions that are held to lend support to a belief in the validity of other beliefs or hypotheses. In the example, the hypothesis refers to the existence of certain living organisms, and the evidence for it would be any number of other beliefs that we hold true, and which make the falsehood of the hypothesis seem highly unlikely.

What then is likely to count as evidence in this example? Such evidence might involve descriptions of specimens observed on the spot, very partial descriptions, of course, selectively composed so as to document the existence of a population possessing features warranting the claim that they constitute a distinct species. This species of frogs will, of course, be identified within some preexisting body of theory or classificatory scheme. Furthermore, our ecologists need to give reasons for their contention that the species is in danger of extinction, and thus they need to provide some sort of evidence that members of this species do not exist in any significant numbers elsewhere on the globe. It is clear that the evidence for the global hypothesis must be at once theoretical and observational, and it should also be obvious that this evidence, far from being a matter of discrete or atomistic facts, involves a complex network of interrelated principles and beliefs based on various experiences, reports, and conjectures. Thus I suggest that the traditional, empiricist versions of foundationalism are wrong, for the epistemic warrant in the example does not derive from some basic class of unrevisable beliefs.

Let us next suppose that one of the parties concerned, the representatives of a firm engaged in the deforestation, makes an unexpected move in the argument with the naturalists. Instead of contesting in any detail the specific arguments, reasons, and evidence presented by the ecologists, they make a different sort of challenge: "This is all very well," they say, "But how can you prove to us with apodictic certainty that the knowledge claims you are making are justified? How do you know that we are not all experiencing an illusion when we think we perceive these frogs? Perhaps we are all being made to experience nonexistent entities by some *malin génie*. Perhaps those creatures are illusions that appear to us by virtue of the transcendental conditions of possibility of our experience, but which do not exist in themselves, whereas all of the other things we experience really do exist in themselves. Our developmental project, which will destroy the forests where those chimerical beasts appear, is only going to rid us of a 'necessary illusion'."

How are our ecologists going to respond to such statements? The opponent has brought forth classical skeptical arguments, challenging the ecologists to provide a proof having to do with a state of affairs that lies, by definition, outside the range of experience. All the previous evidence and reasons for believing in the existence of the tropical creatures are in no specific way contested, but now the belief in the creatures' existence is challenged by a rather different question. How could we know that what the skeptic proposes is impossible? All of the seeming evidence for the creatures being there might be true, but at the same time, the hypothesis of their actually being there might be false. Note that the skeptic is denying, not the evidence or any particular theory, but the claim that the evidence fully justifies (in the sense of "logically entails") our belief that these creatures lie out there at the end of our epistemic chain of reasons and evidence.

It is always tempting at this point to say "who cares," or to add that such doubts are not serious. What could be a better example of "language gone on holiday"—which is how Wittgenstein characterized the sterile war the philosopher wages with the imaginary skeptical opponent? It is indeed wise to think about the place of philosophical arguments in the real world. One way to do this is to focus on the interests that motivate any sudden recourse to such otherworldly arguments. Skepticism in the real world is never as innocent or disinterested as it pretends, the pragmatist might say. Would the commercial developers ever allow themselves to be caught up in such skeptical objections to any of their own claims? Will they undertake to prove, by means of some kind of transcendental argument, the existence of tropical forests, prior to any development? Yet a simple pragmatic reaction does not resolve the epistemological problem, and it is wrong to pretend that only philosophers ever have significant epistemological difficulties. Witness certain debates within literary criticism, where the problems of evidence and justification arise from very concrete and specific inquiries; or consider the hotly contested issues in such domains as medical ethics, where important decisions are intimately connected to conceptual distinctions, the correctness of which is held to be crucial by all parties concerned. Even if we were to decide along with the pragmatists that the only true knowledge is a belief that is really useful, we would still have to ask ourselves how we know what is really useful.

How, then, can we begin to respond to the skeptical arguments? How can we deal with someone who grants the apparent evidence and reasons supporting a belief, yet suddenly imposes exhorbitant requirements on what counts as a telling proof? Can we prove we know anything at all, or is the chain of justifications always left dangling? One of the standard moves at this point is to try to make the skeptical argument destroy itself by turning its own questions around. Roughly, this amounts to asking the skeptics how they think they know that the endangered species is really only a peculiar type of illusion. Surely they can present no grand proof of this belief. As I point out in a moment, this argument is effective in response to certain forms of skepticism. Yet our present skeptic has not really made such an assertion; he is asking us only whether we can fully prove our own contrary belief; in fact, this particular skeptic has no positive thesis to defend at all, not even the idea that there could never be any such thing as a justified belief. He only reports to us about his own *arrhepsia*, his own refusal or inability to decide on what is in principle an undecidable question. Perhaps his argument is best understood as a challenge to us to show how we are justified in pretending to be any different. It is then useless to try and turn this skeptical argument against itself. It does not help to try to tease out the assumptions that underwrite the skeptical challenge, as if the skeptic's implicit reliance upon certain basic norms of reason and evidence could provide the basis for an argument that would rule those exhorbitant questions out of court. In a marvelous passage of *A Treatise of Human Nature*, Hume contends that this kind of argument against skeptics will not work, for it only results in a standoff between reason and the skeptical enemy who has turned reason's own weapons against itself—it is only our nature, he claims, and not reason, that dispells the skeptical doubts and prevents them from having any "considerable influence on our understanding."[5]

Here I feel it is important to acknowledge the necessity of making a concession to our academical skeptic, with whom we agree that it is indeed in principle possible to imagine any number of things. For example, we can conceive, along with Russell, that the world was created only five minutes ago but sprang into existence at that time fully equipped to look much older than it really is, so much so that all of the evidence about the past remains the same. Because the imagined state of affairs may not be a *logical* impossibility, because our reasons and evidence remain the same, there may be no reasonable grounds for refuting such a possibility. Yet we must insist at the same time that this concession is not an admission that the skeptic's fantasy is in fact right. There is no reason whatsoever to believe that any living creature is really a chimera, or that the universe was born five minutes ago, that we are all just brains in a vat, that the supernatural lurks in the night, and so on.

[5]David Hume, *A Treatise of Human Nature* (Oxford: Clarendon, 1978), 186–87.

It is crucial to note that in order to construct his challenge, our skeptic must stipulate that our evidence remains in every respect the same. Given that very major stipulation, it appeared that there is a loophole in the argument of justification. Indeed, the chain of justifications extends only to those beliefs and reasons that we take as evidence for hypotheses about the nature of external reality; it stopped there. There is a network of interrelated beliefs offering local support or justification, but not global Justification by means of which the whole network is plugged into Reality. But what follows from this? As long as we keep this entire network of evidence the same, we can imagine any number of goings-on standing behind it (from chimeras to the Wizard of Oz). But we must not on that basis start to draw any conclusions whatsoever about cases in which the evidentiary network of beliefs is not in every respect the same. So in response to this version of skepticism, the ecologists should make their concession. They may then go on to argue fervently against the deforestation on the grounds that the evidence gives a reasonable justification (if not an absolute, indefeasible proof) of the belief that there is an endangered species out there—and maybe even a very special chimerical beast. Should not the violent destruction of such creatures be prohibited?

Inquiry is usually concerned with cases where the evidence for and against different hypotheses is not precisely the same, and consequently the kinds of justification problems that are encountered differ from the one we just discussed. There are sometimes different reasons for concluding in favor of one belief as opposed to another, reasons having to do with the logical norms, empirical theses, and other evidence characteristic of the arguments under consideration. This kind of justificational problem is often simplified by the many norms and assumptions that remain constant, providing a stable background against which certain differences become telling. To return to my example in order to illustrate that point, we can usually expect that the two parties in our dispute clearly have many beliefs in common, and this in spite of an opposition to each other that is likely to be very bitter. For example, they share the idea that whatever the facts turn out to be, one can say in advance with perfect certainty that either the creatures in question exist out there in the jungle or they do not.

Now let us again suppose that our commercial developers try something that would be unusual in such arguments. And so they argue: "You know that species of frogs that you've been worried about? Well you shouldn't worry, because it's like this: it's a matter of interpretation. Strictly speaking, we can say that they don't exist and that they do exist." When we are confronted with such statements (and in literary criticism we sometimes are), we may wonder what kind of claim is being made. Is it a matter of some radical rejection of certain norms of argumentation, norms that could be challenged as arbitrary or unfounded? Is it the principle of noncontradiction and/or the law of the excluded middle that is at stake here? So we ask our interlocuter how the existence–nonexistence of these frogs reveals itself. Do these frogs pop in and out of existence in rapid succession, or is it true once and for all that they both exist and do not exist? If it is the former, then this unusual claim can be deemed an empirical question, by which we mean only that there is in principle some kind of reasoning or evidence that could bear upon the issue—probably the same kind of evidence that would be pertinent in relation to other claims about the existence of other tropical animals. Even if this is an empirical claim, it would be unreasonable to take it very seriously—it would be strange to conduct extensive observations to see whether the frogs really display such astounding behavior because it would be a matter of applying an empirical and scientific method of testing to a case that flies in the face of a well-tested set of expectations about the facts of the natural world. This is not to say that these expectations are held to be infallible; it would be better to say that there are too many real questions left open by them to become concerned about one rather improbable report. Although we are not absolutely certain that the frogs do not oscillate between being and nonbeing, we can be certain, beyond all reasonable doubt, that they do not.

If what our interlocuter has in mind is the second kind of claim (there is a *Rana paradoxa* that always does and does not exist), then it is a different matter, for we are warranted in judging this belief incoherent in relation to an implicit norm that excludes the assertion of *p* and not-*p*, *in the same sense and at the same time,* from what we can call reasonable argumentation. Whenever someone tells me that they do not in fact "believe in" or care to respect such a norm, I like to see how far they are willing to go. Are they, in trying to argue with me over a supposedly arbitrary principle of noncontradiction, still working with it? Are they willing to say that this principle both holds and does not hold, at the same time and in the same sense? When they go on insisting that it does not hold, I ask them why they are still applying the principle; mustn't they also agree that it holds? Otherwise they have only made one contradictory statement, which can be identified as an exception to the principle—and which makes sense by virtue of there being a level at which the principle is inviolate.[6] Surely they cannot be content with only one exception to the abhorrent principle, which does not stipulate, in any case, that no one could ever say something contradictory. Yet should they attempt to enact some kind of global violation of the principle, this leads to a regress: for every contradiction that they agree to state, for example, "*p* and not-*p*," there is yet another proposition to be admitted—"(*p* and not-*p*) *and* not-(*p* and not-*p*)." And so on. It seems, as Aristotle suggested, that the basic intuitions behind such logical norms provide an implicit principle or *arche* at work in making intelligible even those statements which, at another level, are supposed to be subverting them.[7] This would not be a framework one could choose to adopt or not to adopt; rather, it would be *the* conceptual schema within which local choices—be they coherent or incoherent—are made. That kind of framework would not be put in question when someone utters now *p* and *then* not-*p*. Moreover, it is difficult to conceive of what evidence could possibly lead us to doubt its applicability to the realities of the natural world; can we really conceive of what it would mean for a living creature to exist and not exist, at the same time and in the same sense?

As I would not like the reader to think that the object of this section is a criticism of an attitude that is but the product of my own imagination, I feel obliged to give at least a single example of the sort of claim under discussion. In her preface to the English translation of Jacques Derrida's *De la grammatologie*, Gayatri Spivak castigates with a certain arrogance the "usual superficial criticism" of Derrida that complains that he "says he is questioning the value of 'truth', and 'logic', yet he uses logic to demonstrate the truth of his own arguments!" To attack Derrida in this manner, Spivak contends, is "to overlook the invisible erasure," to fail to comprehend that Derrida makes the "problematic" status of his own terms his "overt concern." Hence we are given to understand that special rules must govern any profound discussion of Derrida. The central issue becomes that of assessing the crucial strategy of erasure, namely, Derrida's former gesture of employing a term only to cross it out. Spivak

[6]My argument here echoes Jean-Pierre Dupuy's analogous critique of the epistemology of pure paradox in "La Simplicité de la complexité," in *Ordres et désordres* (Paris: Seuil, 1982), 211–51.

[7]Aristotle, *Metaphysics*, bk. 3; for an unsympathetic commentary, see Dennis Rohatyn, "Aristotle and the Limits of Philosophic Proof," *Nature and System*, 4 (1982), 77–86; see also Karl-Otto Apel, "The Problem of Philosophical Fundamental-Grounding in Light of a Transcendental Pragmatic of Language," *Man and World*, 8 (1975), 239–75.

herself moves on—perhaps too hastily—to discuss the meanings of this procedure, adding, however, a little caution by using it herself: "Derrida's ~~trace~~ is the mark of the absence of a presence, an always already absent presence, of the lack at the origin that is the condition of thought and experience. . . . At once inside and outside a certain Hegelian and Heideggerian tradition, Derrida, then, is asking us to change certain habits of mind: the authority of the text is provisional, the origin is a trace; contradicting logic, we must learn to use and erase our language at the same time." Thus the putative necessity of a term for the origin and essence of the sign must be recognized but must also be undone, for the origin and essence are not really such but are traces of disappearance and nonpresence. What is this process of erasure, then? Spivak answers that it is a matter of changing certain habits of mind, and it certainly appears that those habits have to do with logical norms of identity and noncontradiction. Spivak adds that Derrida says of his own concept of erasure that it is "in fact contradictory and not acceptable within the logic of identity."[8] To complain, then, that Derrida uses logic against itself may be a superficial attack, for he certainly makes no claim that logic is left intact or that the concepts of "trace" or "erasure" are acceptable in such terms. On the contrary. But one step is obscured here. If a superficial attack is one that pretends that logic remains intact, a superficial defense is one that pretends the contrary, imagining that we can really change those mental habits once and for all. The full playing out of the undecidable requires saying that we can *and* that we cannot; it will not do to assert univocally that the "lack at the origin . . . *is* the condition of thought and experience." We must also add that the lack at the origin, which is a certain nonlack of the origin, *is not* the condition of thought and experience, which is also nonthought and nonexperience. In this way we remain (and do not remain) profoundly faithful to Derrida, for in his terms (which are not his terms) the logical "habits of mind" are at once *necessary* and *under erasure.*

In regard to the status of logic and paradox, it does not seem correct to agree with the speculations of Lucien Lévy-Bruhl, who attributes paradoxical propensities to what he calls "the primitive mentality."[9] According to Lévy-Bruhl, the primitive mentality frequently denies the rules of rational thought, rules that Lévy-Bruhl

[8]Jacques Derrida, *Of Grammatology*, trans. Gayatri Spivak (Baltimore: Johns Hopkins University Press, 1976), xvii–xviii.

[9]Lucien Lévy-Bruhl, *Les Carnets* (Paris: Presses Universitaires de France, 1949).

equates with a somewhat vague version of Kant's categories and a priori forms of intuition. One of his examples is that of native beliefs whereby someone is supposed to be at two places at the same time. My response to this and similar examples is to ask whether the theorist has really grasped the sense of the beliefs in question. It seems very likely that a person can be thought to be in two places at the same time—but not in the same sense. In the kinds of examples documented by Mircea Eliade, a shaman is beneath his hut, performing a ritual dance, and at the same time his spirit has gone to an invisible domain where it does battle with the forces that are thought to have caused his patient harm.[10] Clearly there are at work here some particular assumptions about the identity of the shaman, assumptions about the dual nature of his person as well as of the world in which he lives, which is at once sacred and profane, invisible and visible. Yet the existence of these assumptions does not corroborate the idea of a logical error or contradiction at the heart of the mythical participation; the sacred and the profane, although they stand in opposition to each other, are typically thought to be complementary and not in contradiction. It is particularly wrong to attribute such nuances of meaning to "the natives" or to "primitive thought" alone, if by this is meant some vast group of non-occidental cultures (or worse, races). The most rational-minded Western individual engages in these so-called participations on a regular basis without transgressing logic, particularly in relation to the kinds of presence-in-absence involved in representations. We may believe that we saw Arletty in a film; yet we also believe that she is not "really" in the film. If someone asks whether she was in *Hôtel du Nord* or not, the correct answer is yes. What is excluded from "our framework" is the possibility of it being reasonable to say that we both saw and did not see Arletty—in precisely the same sense—when we saw the film.

My discussion of the epistemological problem of justifying beliefs has so far had little to do with specifically scientific knowledge because it has been necessary to discuss certain kinds of skeptical objections that challenge much more basic epistemic norms, norms having to do with the justification of beliefs as such. It would be possible, of course, to go on in a similar vein for several volumes, and indeed, epistemologists are still attempting definitive refutations of the whole battery of skeptical arguments, just as skeptics are working on their counterstrategies. Even so, I hope to have accomplished a few things in my cursory discussion of these issues. More precisely, I

[10]Mircea Eliade, *Le Chamanisme* (Paris: Payot, 1968).

think that it has been shown that skepticism about the possibility of true beliefs about the external world is always possible, but that it works within limits that sharply reduce the potential range of validity (and even more, the interest) of such challenges. The most radical forms of skepticism, that is, those that represent the most serious challenge, are coherent, then, but they find purchase only at the cost of leaving everything else in place. Their skeptical challenge seems to be important only if we add one additional premise: it is reasonable to consider a belief to be knowledge only when we know with absolutely certainty that it is true. But this additional assumption needs to be argued, for it expresses only one of our options concerning what we may deem sufficient evidence for the justification of a belief. It should be noted, for example, that it is possible that many of our beliefs could truly correspond to what is the case without our having this kind of absolute evidence or justification for them. Yet our skeptic prefers an absolute theory of justification that leads us to sacrifice all of these beliefs to the possibility of doubt; fear of error and a desire for absolute reliability are given first priority.

Other forms of skepticism seem to arise from *within* a network of evidence that implicitly includes possibilities for justification, thus already allowing for a variety of arguments in response to the radical skeptical challenge. When these presuppositions about evidence and justification flatly contradict the skeptical propositions and questions, we have an incoherent skepticism. This latter type of skepticism, then, is self-destroying, for if it tries to challenge such basic norms of reasoning as induction, it can easily be shown that the very challenge relies upon these same norms.

As a final illustration of the first type of skepticism (that which installs itself in the place of transcendental foundations), it may be interesting to refer to one of Jacques Derrida's argumentations, which is in many ways characteristic of a certain vein of his work.[11]

[11] I make no attempt in this book to provide a comprehensive or detailed evaluation of the epistemological positions of Jacques Derrida, nor of those of so-called deconstruction. Derrida's writings are obviously very brilliant, yet I disagree with what I view as an overall tendency toward a kind of skepticism, as well as a habit of engaging in elaborate pseudodemonstrations of impossibility, "argued" by means of lengthy close readings of a selective philosophical corpus. I also disagree profoundly with Derrida's insistence on the priority of negative theology over other forms of knowledge and inquiry. Finally, I see no justification for his massive investment in the philosophy of Heidegger, whose works he has recently called "at once terribly dangerous and madly *drôle*" (in "De l'esprit," lecture at the College Internationale de Philosophie, March 14, 1987). On the subject of so-called deconstructive criticism, or of those North American new critical appropriations of Derrida's work, I agree with those who deem such work trivial when it amounts to a change in the set of themes,

In a discussion of concepts of "authority" and "signature," Derrida notes that signatures are meant to serve as a trace marking the single "event" of the author's presence to his own writing. Yet this presence is only an effect; the author is in fact absent from the finished text of his or her signature—a signature can be falsified, for example. If the signature were really to be attached to its source, or to the author's particular act, it would have to retain what Derrida calls "the absolute singularity of the event and form of signature: the pure reproducibility of a pure event."[12] Derrida's rhetorical question at this point is whether such a thing as a signature is possible, and his response is "yes, of course, every day." Yet this is hardly an important concession for Derrida, who goes on in the same passage to add his objection, an objection that strikes me as a brilliant example of the classical variety of academical skepticism: "But the conditions of possibility of these effects is simultaneously, once more, the condition of their impossibility, of the impossibility of their rigorous purity." If one is in a hurry, this text reads as if it were asserting something rather astounding—that the conditions that make the practice of signatures possible are also the conditions of their impossibility. That would surely be the end, not only of checkbooks, but of our logical habits of thought. To read more attentively, however, is to note that this first impression is erroneous. All that is really impossible is the "rigorous purity" of the signature. Why would this rigorous purity be impossible? Because the conditions that make the use of signatures possible involve the repetition of the signature's form, its iterability and sameness—which are held to be the opposite of the act of inscription as a purely singular event. (But are tokens the logical *opposites* of types?) Perhaps one finds the passage misleading at first glance because of the way its appositional structure evokes an equation or identity, when in fact what is designated on the two sides of the comma are very different notions. The text reads as follows: ". . . the condition of their impossibility, of the impossibility of their rigorous purity." The left side of this apposition refers to what I have spoken of above as the world of arguments and evidence, and it is in this world that authentic signatures and forgeries have their place. This is the world in which the conditions of possibility of signatures are in no way the conditions of their impossibility—if only because

terms, and typical moves used to produce readings of particular literary works. For further argumentation on Derrida and related issues, see Chapters 5 through 7.

[12]Derrida, "Signature événement contexte," in *Marges de la philosophie* (Paris: Minuit, 1972), 391, see also 367–93.

signatures, which Derrida strangely refers to as "signature effects," are not impossible at all, a fact that Derrida clearly acknowledges: "Signature effects are the most common thing in the world." But the conditions of the *impossibility* of signatures to which Derrida "simultaneously" refers are not of this same world; they belong to a realm of "rigorous purity," a realm evoked when one tries to imagine the existence of such things as absolute singularities and pure events, and where these singularities and events are nonetheless brought back into the other world to be confronted with the semiotic practice of making particular inscriptions that are recognized and grouped as iterable types or classes. It is this juxtaposition of the idea of the singularity of the trace with its identification as a mere token of a type which creates a sense of incongruity and paradox. One point should be stressed: in the world of evidence and arguments, the everyday world where signatures are clearly possible, the author's signing of his or her name is not a pure event or an absolute singularity, and it makes no sense at all to look for such an event or singularity at the level of analysis pertinent to the discussion of the symbolic practice of signing one's name; to sign competently at all is to create a mark that is a token of a type of marks; it is *not* to engage in a singular event having a rigorous purity. The same essential claim would be valid should the analysis be displaced to the level of the act of physically inscribing a mark on a piece of paper, for here as well the absolute singularity of the act or event is not pertinent insofar as the questions of a nonskeptical inquiry concern types and kinds of events and not absolutely nonreproducible features of particular entities.

To read this passage rigorously, then, it is necessary to keep the two sides of Derrida's apposition clearly in mind, and to observe the ways their juxtaposition creates a certain effect. Nothing indeed—be it a matter of argument or evidence—can ever ground signifying practices—which necessarily rely upon iterability and the type–token opposition—in some realm of rigorous purity and absolutely singular events. But once we acknowledge this, absolutely nothing follows for the world of signatures and forgeries, of checks and author's rights, of attributions and citations. Nothing follows for this world, which is the world of the "work and fact" of signification and meaning and not the world of its transcendental conditions of possibility, of its absolute origins and nonorigins.[13] And one should

[13]Derrida, *Of Grammatology*, 63.

wonder in the name of what transcendent domain one would find it necessary to refer to the world of signatures as a mere effect.

Framework Relativism

As I mentioned before, such transcendental skepticisms, although they may be perfectly coherent and wholly irrefutable, make no difference whatsover as long as we do not mistake their questions for valid statements about the world of arguments and evidence. That kind of assertion is the province of a second type of skepticism, one that is incoherent and self-destroying. An example of this latter type of skepticism is the set of stances I refer to as framework relativism. This position on the epistemological problem, which stresses the relativity of all truth, exercizes a broad appeal, for it seems to be supported by compelling observations. Moreover, it advocates what seems like a modest and even ethical pluralism. Indeed, the framework relativists find their starting point in the shortcomings of many of the classical epistemological arguments. It is possible to show, for example, that the previous attempts to give knowledge a rigorous foundation were marked by an ahistorical bias that led to a focus on isolated bits of knowledge having to do with fairly arcane examples, such as whether I can know that I am not dreaming. What was left out in such deliberations was the complex texture of knowledge, its historical dimension as well as its existence as a network of interconnecting beliefs and practices. Related to this shortcoming was the individualistic bias of such approaches, for the subject was imagined to be alone in *his* quest for certainty.

The framework relativist's argument finds further support in something that it seems unreasonable to deny, that is, the observation that different groups of people have radically different beliefs about the world. There *are* many different cultures; some people believe in witches and flying carpets, others believe in electrons, gluons, and who knows what else. Even if we somehow were to pick out only those conceptions having to do with something called "nature" (that is, something that *one* such framework has isolated as such), thereby ignoring the incredible relativity of norms, mores, values, visions of history, and so on, there is still no basic agreement to be observed. "Fire burns here and in Persia," Aristotle said[14]—but

[14]Aristotle, *Nicomachean Ethics*, 5.7.1134b–1135a.

is it true that the same thing is meant within the different cultures? Has not Aristotle already done some "translating" in order to be able to make this assertion? Not only must we recognize, then, that knowing subjects and their beliefs are possible only within frameworks; we must further admit the existence of a multiplicity of such frameworks.

It is at this point that the framework relativists take a step that does not directly follow from what has already been said, for they go on to contend that the existence of different belief systems entails the impossibility of justifiably asserting that any one conception (about nature or any other circumscribed region of the real) is more true or accurate than any of the other conceptions. In the framework relativist's view, "truth" equals "true relative to a framework." Moreover, for this thesis to have any real force, it must also be the case that these frameworks are radically divergent, that is, mutually contradictory. For some significant and basic set of beliefs, there must exist at least one framework in which they are held true, and at least one other in which they are held false. In other words, there exist, across a non-unitary time and an unintegrated space, a plurality of different "knowledge-networks," no one of which can pretend to stand as the axiomatic and authoritative metalanguage for the others, in any way whatsoever.[15] But what must be assumed in order for this to follow? Several important assumptions are at work here. At one extreme, there is some version of the incoherent consensus theory of truth that we earlier discussed. There are also some interesting, yet unspecified and dubious, metaphysical and ontological conceptions as well—perhaps an idealism whereby mind or language is thought to cause or create all reality; perhaps an implicit dualism whereby the ultimate reality is a kind of putty that is carved up or organized by different cultural schemata. The details do not matter; the main point is that far from being the lovely and generous pluralism that it pretends to be (in opposition to the violent monologism of ethnocentric science), this framework relativism is at bottom a particular set of assertions about what does and does not exist, about what it is accurate and inaccurate to assert, about what can and cannot be justified. In short, framework relativism is a particular theory of truth. The validity of this set of assertions is not obvious, it

[15]Such appears to be one of the tenets of postmodernism as elucidated by Jean-François Lyotard in his *La Condition postmoderne: Rapport sur le savoir* (Paris: Minuit, 1979); *Le Différend* (Paris: Minuit, 1983); *Tombeau de l'intellectuel et autres papiers* (Paris: Galilée, 1984); and *Le Postmoderne expliqué aux enfants* (Paris: Galilée, 1986).

cannot be derived simply from the fact of cultural diversity. Just how diverse is this diversity? one must still ask. (I do not pretend here and now to bludgeon the framework relativists for a final answer to *that* immensely difficult question; rather, they are the ones who implicitly claim to have settled this matter when they assert that cultures and worldviews exist as unique, self-enclosed, radically incommensurable and heterogeneous systems.)

The key to the flaw in the framework relativist's position can be identified by broaching a simple question: what is the framework relativist arguing with us over? The answer to that question can, at the most general level, be put in this manner: over what is the case. Here we can perceive the difference between framework relativism and the kind of skepticism discussed earlier; the skeptic argued with us not over any particular features of the world of arguments and evidence, but over imaginable possibilities that do not run contrary to anything that we take to be part of the world of our reasons and evidence. Nor did the skeptic hope to prove to us that these strange possibilities were actual realities, only that they were possibilities that we were not able to rule out with absolute certitude. The framework relativists, however, make assertions about the world of arguments and evidence, assertions that presumably make a difference within this world. Hence they share the assumption or attitude that is a basis of all reasonable inquiry and dialogue, namely, the attitude whereby the goal is to try to determine what is the case. The framework relativist may present himself as a heroic pragmatist who only wants to tell stories about what he likes and dislikes; he may speak only in terms of how he would prefer to have us use one metaphor rather than another; along the way, he will add that the metaphors of truth, evidence, and mirrors are shopworn, tired, and no longer useful.[16] This is the moment when we perceive that the basic attitude these tired metaphors are supposed to express is still alive and at work in the pragmatic hero's own storytelling. His statements have the implicit aim of convincing us that it is true that we can stop worrying about the truth and only tell stories; his story asks us to believe that it is correct to think that the metaphors of truth are old hat, truly useless, wrong. What is asserted is that the constitutive desire of philosophy—the desire not merely to repeat and obey the dogmas of mere opinion but to search for knowledge—is in fact a hopeless chimera. In spite of all of these disclaimers, then, when we

[16]My reference here is to Richard Rorty as he presents his philosophical hero in "Texts and Lumps," *New Literary History*, 42 (1985), 1–16.

are through listening, what rings in our ears is still "it is the case that." And this metaassertion, implicit throughout the framework relativist's discourse, contradicts the various individual claims that discourse can never refer to the fact of the matter.

At a much more specific level, it is often possible to point out that other, much more particular rational norms of argumentation and evidence are being relied upon as well. There is an attempt at coherence in the chain of reasoning; facts and documents are cited, examples are given; rhetorical devices are used to enhance and make clear the strong points of the argument. Authorities may be cited for the sake of persuasion, but we have not yet encountered in the literature of the framework relativists any explicit arguments to the effect that cultural relativism is true because some charismatic leader or cult authority ordains it so. Perhaps this is implied when they tell us that institutions, professions, and traditions are the final authorities and that we should forget about truth and values for this reason. Yet we are asked to agree because "it is the case" that there is no source of truth or legitimacy outside the fact of social authority. Clearly, the idea that beliefs are often legitimated through social authority alone is an indispensable insight for any descriptive sociology of knowledge. In epistemology, however, its status is rather different, for here the simple authority of an individual cannot serve as the end of the chain; it must itself be given a further justification. One may very well recognize the existence of experts whose opinions are most likely to be valid, but it is not solely by virtue of social recognition that these beliefs would in fact be justified. If someone is an expert, it is because he or she possesses an effective expertise about something, and the reliability of this expertise can be challenged and checked by others. Knowledge, then, is not reducible to the social fact of the existence of what Jacques Lacan calls *sujets supposés savoir,* and the framework relativist's reliance upon such notions is no ground for a compelling argument.

Framework relativism is incoherent, then, at a variety of levels; it represents a contradictory blending of skeptical doubts about the possibility of justified and true belief with assertions about the nature of all knowledge claims. Thus the framework relativist's first major premise is that "true" should always be transcribed as "true relative to a particular framework," a basically ambiguous and confusing expression. The second step is to assert that, for some basic beliefs (e.g., those about knowledge and nature), there must be radically different frameworks where these beliefs are true relative to at least one, but false relative to another. When combined, these two assertions give rise to the incoherent and absurd nature of the position, for it follows, first of all, that the framework relativity of truth is itself relative to a framework, and also that the belief in the existence of radically divergent frameworks is also only true relative to a framework. This tissue of confusions often leads the framework relativist to try to diminish the scope of the assertive half of the position by saying that no ontological theses are made concerning whatever may or may not exist outside "our" own framework. This relativist is not saying that a variety of frameworks really exists, only that given our own framework, we have to conclude that framework relativism is correct; outside it, or within a different framework, it could be wrong, and there could, for example, be justified truths about nature. But is this version of framework relativism any more coherent? One wonders what it is that is *not* included in "our" framework, a framework that was, after all, defined as the englobing and fully autonomous network of concepts, norms, and experiences organizing our cognition and determining our truths. If we are really locked into such a framework, as this view contended, how is it that we are now permitted to imagine the opposite possibility? How is it that we are now allowed to compare our limited framework to other possible frameworks, even ones within which framework relativism turns out to be illusory? The strongly systematic, enclosed, and binding characteristics of the framework, characteristics upon which this relativism heavily relies in building its antifoundational arguments, are now being loosened so as to restore a semblance of coherence, yet this is achieved at the cost of depriving the position of its basic thesis, which was that only framework-relative truth is possible. When it is presented in this revised version, we can only agree with framework relativism, for we indeed believe that other kinds of truth are possible and that framework relativism is an erroneous theory of truth.

To sum up, we have seen that framework relativism is either incoherent or trivial. It is incoherent when it asserts that truth always or necessarily comes in limited and incommensurable frameworks (and hence is limited and contingent truth), for this very assertion is an unlimited thesis about what is necessarily the case about knowledge and reality. When framework relativism avoids this contradiction, it typically reverts to saying that if you happen to see the world from its perspective, then you see the world from that perspective—which is trivial. Whenever framework relativists suggest that knowledge and reality are that way, or that anyone must see them from their perspective, the contradiction returns. The kernel of truth in this position is the insistence that all beliefs are interconnected and that single knowledge claims and doubts emerge against a background or network of beliefs, assumptions, and principles. This is an invaluable insight in a theory of cognition, but it is wrongly presented as the decisive element in a theory of truth in which the vague notion of "true relative to a framework" wrongly replaces "true relative to what is the case." In other words, we agree that all knowledge is made possible by a particular network of beliefs and assumptions, but it does not follow that the validity or truth of knowledge claims is only a matter of the internal relations within a framework of beliefs. The truth of beliefs is not holistic, or solely a matter of the internal relations within some purely self-enclosed cognitive system, but a matter of their individual and collective relation to something else. Perhaps this is what Bertrand Russell has in mind when he writes that the condition of the truth of a belief is not in the belief or mind, but in the state of affairs that the belief is about; beliefs depend on the activity of minds for their existence, but not for their truth: "Minds do not *create* truth or falsehood. They create beliefs, but when once the beliefs are created, the mind cannot make them true or false, except in the special case where they concern future things which are within the power of the person believing, such as catching trains. What makes a belief true is a *fact*, and this fact does not (except in exceptional cases) in any way involve the mind of the person who has the belief."[17]

At this point the limited interest and validity of the kinds of skepticism under discussion should be fairly clear. What I have said in regard to skepticism, however, hardly amounts to an adequate presentation of a positive conception of truth and of justification—even less a comprehensive theory of knowledge—and the reader may feel that the chain is still dangling in the air. Let me briefly restate the problem. If we have a belief about what is the case and want to know if we are justified in holding this belief, we need to know what really is the case to find out if our belief matches it. How are we going to get some additional confirmation of our belief? If we need another belief (e.g., one about evidence) for this purpose, we seem to be led into a regress. The key to the problem is the difficulty of specifying how it is that we can be sure that beliefs match or link up with the facts. Another way to put this is say that it seems that the theory of truth is the pressing issue. The central problem is not the regress or

[17]Bertrand Russell, *The Problems of Philosophy* (Oxford: Oxford University Press, 1912), 129–30.

circle of beliefs standing in relation to other beliefs, but assurance that there is a substantive, external, or nonepistemic source of truth for knowledge, or for the criteria and methods of knowledge in question. Yet if the epistemological inquiry is to be pursued without interruption of the chain of justifications, it must be recalled that this supposed nonepistemic source of truth is going to have to be specified in epistemic and justificational terms, that is, in terms of beliefs, propositions, concepts, percepts, intentional states, idealized rational acceptability, or what have you. In other words, the theory of truth must still be backed by a nonregressive account of justification. Hence the very terms of the discussion dictate that there is a basic ambivalence surrounding what could be called the "evidence." Evidence must play a double role, for it is at once within the epistemic chain and outside it; it is part of the chain of beliefs, yet it is also the special link anchoring that chain of beliefs to something that is held to be transcendent in relation to it, such as a factual, objective nature that in no way depends on belief for its existence. For a belief to be properly grounded, then, would be for it to have some kind of basis or support in reality, understood as the external, nonepistemic state of affairs with which it would be somehow linked. Yet how can a belief be directly linked to facts or reality without ceasing to be a belief? Would that require us to hold that the belief partakes of the reality in some sense? And what is the nature of the bearer of this truth which achieves such a link to something outside it? What about the belief that refers to my true belief? Not only must I know, but I must know that I know. It is the ambivalent status, the dual nature that is required of the justifying belief, proposition, or state of mind, that makes it so difficult to formulate an adequate response to these problems.

If this problem is to be solved, there must be starting points, that is, a belief or group of beliefs that stand as evidence and that can be used in the evaluation of other hypotheses about what is the case. But what makes a particular belief, concept, or proposition serve as evidence? In pointing in the direction of a possible justification of belief by means of evidence (in opposition to arbitrary authority), I have already indicated what I take to be the crucial starting point for any correct account of reasonable belief: it is reasonable to conclude in favor of the belief for which there is sufficient evidence. A theory of justification, then, amounts to a specification of this latter concept. It should be clear that the word "evidence" is no magical charm, capable of quickly guaranteeing the adequacy or truth of anyone's beliefs about the natural universe. How does evidence resolve the problem of the ambivalence of the justifying instance discussed above? I have already sounded a warning against hasty assumptions about the nature of evidence—such as the foundational strategies of naive empiricism, which hope to begin with indubitable bits of ordinary, that is, concept-free, "perception," or purified units of "sense data"—and other similar philosophical abstractions. Nor does it seem promising to start with a purely rationalistic derivation of some Indubitable Proposition—or "that without which I cannot think"—such as the self-grounding thought of the existence of the doubting consciousness (more on this line in Chapter 4). Moreover, I have briefly suggested that the centrality of the problem of error militates against the development of anything like a necessary and universal framework of certain and justified knowledge. People's beliefs may have to do with faulty appearances and phantasmic projections a lot of the time, and the epistemologist goes wrong in trying to get their beliefs to link up with reality in some regular and automatic manner; such a linkage—that is, a fully infallible one—must be understood as being at best normative, not descriptive. All that we should try to *describe* is a fallible yet sometimes accurate link between beliefs and facts, although we may highly prize the goal of a much less fallible model. As we saw above, the radical skeptic's theory of justification is one requiring total reliability, which in effect means that no claims can ever be justified.

It has already been indicated that beliefs are justified in terms of reasons and evidence, and it is the role of epistemology to discuss the merits and demerits of different ways of construing these things. Reasons and evidence must in turn be related to other deep-seated and basic assumptions, for example, a theory of truth. I have suggested that framework relativism is an epistemology crippled by its incoherent theory of truth. One may usefully oppose to this theory the barest and most minimal realist principle following which the objects of inquiry (which need not be held in advance to be material or discrete, etc.) are states of affairs as they exist independently of the act of inquiry. Of course it is possible here to tinker at great length with the specifics of the formulation; for example, what sort of independence do we have in mind here—logical, causal, ontic, epistemic? Surely we do not want to say that the objects of inquiry are wholly independent of any epistemic relation with any mind, for then they could not be known. And what if the object of inquiry is some intentional or mental state—that of someone else, or one's own? In this case their mind-independence must have to do with a certain binding, intersubjective status and not with an ontological autonomy (the realism under consideration here is not a scholastic realism whereby numbers and universals exist independently of the mind). These matters are very complicated; at this point in my presentation, a very general idea of genetic, productive, or causal independence suffices to give an idea of the more general and intuitive point being made. What matters is that the truth of a belief cannot be the *sole* product of the belief itself, nor of the belief's relation to other claims and assumptions. The belief is not true because it is believed; it is true because the state of affairs to which it refers is in fact the case. The state of affairs was not created, produced, or "overdetermined" by the act of inquiry; successful inquiry does not merely discover something that is only and entirely the consequence or product of its own act of discovery. Another way to put this is to say that the kind of truth that I have in mind is not autopoietic; what we are after are discoveries, not inventions. A true belief refers to an actual or factual state of affairs, it does not create it, and the Protagorean relativism in which frameworks of beliefs create all reality is incorrect.

What has just been said does not characterize in any detail the specific nature of the referential link whereby the chain of reasonings and justifications is connected to what is the case. Such a link is assured in a variety of ways, and it is not my aim in this chapter to offer a more detailed and systematic account of the notions of truth, sufficient evidence, and justification. Quite generally, it may be stated that in genuine cases of inquiry, the refinement, justification, and invalidation of beliefs or hypotheses are made possible by means of a network of different cognitive practices, which are intersubjectively testable and iterable. These practices include such activities as theorizing, measuring, perceiving, detecting, experimenting, observing, and inferring. I suggest more on these matters in Chapter 3.

In the discussion of natural science that follows, I remain committed to the very general realist norms of inquiry just touched upon. It follows that the best account of science is the conception that best satisfies the goal specified by these norms (and please note that this *could* be an account in which science is shown to be an extremely irrational enterprise). I do not pretend to have presented anything like an exhaustive or technically rigorous description and definition of the basic epistemic principles in question. I alluded above, for example, to certain logical norms supplying an implicit and guiding framework of rational argumentation and inquiry, but I did not provide a particular description of these norms or situate them ontologically.[18] Nor have I said anything in detail about the kinds of communicational and intersubjective norms that guide and make possible a knowledge that approaches the goal of justified and true belief. Although I have offered the minimal suggestion that truth involves a belief's being linked to the facts, I have provided neither a rigorous analysis of this referential link nor a single, binding definition of such crucial terms as "justification," "evidence," and "knowledge."[19] I do think that I have shown the limits and weaknesses of the arguments typically used to reject, in an a priori manner, the possibility of justifying knowledge.

It should be pointed out that the very broad principles of inquiry that have been discussed—and that have been situated in relation to some skeptical alternatives—owe little to the specifics of the modern natural sciences. On the contrary, it seems more accurate to say that the broad notion of rational inquiry in question is not at all scientific—even if we would eventually go on to say that scientific knowledge is a result of an intensified application and development of this same kind of inquiry, and that the general idea of a scientific method amounts to an explicit systematization of its basic principles (but this is not to be taken as an adhesion to the view of science as just organized common sense). This does not mean that the results are qualitatively the same, or even that the worldview of successful prescientific inquiry and the successive metaphysical "world pictures" of the modern sciences are somehow at base equivalent. Far from it. Yet to contend, as does Edmund Husserl, that there is a profound and mysterious rupture between the prescientific modes of prediction, induction, hypothesis formation, and causal explanation, on the one hand, and the ideal of "calculation" that guides the new "mathematical physics," on the other, seems to be an exaggeration, an exaggeration based on an overly positivistic (and hence anachronistic) understanding of Galileo's achievement.[20] Indeed, it seems

[18]On this problem, see Peter Gärdenfors, "The Dynamics of Belief as a Basis for Logic," *British Journal for the Philosophy of Science*, 35 (1984), 1–10; Fred Sommers, *The Logic of Natural Language* (New York: Oxford University Press, 1982).

[19]A useful source on the question of evidence is Peter Achinstein, "Concepts of Evidence," *Mind*, 87 (1978), 22–45; see also Achinstein, *The Nature of Explanation* (New York: Oxford University Press, 1983). Readers who would like a more technical account of justification are referred to the article by Rosalind Simson cited above, "An Internalist View of the Epistemic Regress Problem," which points to a promising alternative to the false choice of foundationalism, coherentism, and contextualism.

[20]Edmund Husserl, *The Crisis of European Sciences and Transcendental Philosophy: An Introduction to Phenomenological Philosophy*, trans. David Carr (Evanston, Ill.: Northwestern University Press, 1970), 31–32.

more pertinent to point out that it is far from obvious that it was first of all a mathematical physics that established the concept of a "nature" having knowable and determinate regularities or laws—even if it was the success of the modern sciences that led to the radicalization and hegemony of such a viewpoint. In any case, my procedure here is not that of giving myself the "fact" of scientific knowledge, taken as the ideal, and then proceeding to develop the epistemology that would best fit it. Whether there is such a thing as scientific rationality, what the status of its knowledge claims really is, and whether an epistemic demarcation between science and nonscience can be justified—these are the vital issues for an inquiry into science that cannot give itself in advance any specific technique or procedure of empirical scientific research. Yet what this inquiry does rely upon in advance are some basic and minimal norms having to do with the difference between coherent and well-supported accounts, on the one hand, and wholly speculative and incoherent fantasies, on the other.

On the Sociological Turn

A major challenge to epistemology has arisen from a sociological perspective that promises to supplant both the issues and approaches of epistemology. In regard to the natural sciences, the foremost issue that this "sociological turn" puts in question is that of the epistemological justification of scientific knowledge. Indeed, the supposed centrality of the demarcation problem, a centrality taken for granted by positivist philosophies of science, is today a major point of contention.[21] Thus I consider at some length recent tendencies to reject or ignore the demarcation problem altogether. For what is at stake is the kernel of the proof given for an historicist account of natural science.

It seems crucial to recognize from the outset that the old position, following which the problem of science's epistemological justification was held to be *the only* key to understanding science, is quite

[21]Unfortunately, in literary and humanistic circles, the word "positivist" has only one essential meaning: "reprehensible and erroneous viewpoint in favor of science." This view is sheer ignorance, not because none of the various positivisms is neither reprehensible nor erroneous, but because such a usage amounts to an erroneous and reprehensible flattening of history. For an excellent discussion of the varieties of positivism and its critics, see Norman Stockman, *Anti-Positivist Theories of the Sciences: Critical Rationalism, Critical Theory, and Scientific Realism* (Dordrecht: Reidel, 1983).

simply wrong; science is a multifaceted reality of interest to people in many different ways. Some of these interests may simply be irreducible to each other. To take an interest in the epistemological questions raised by the existence of science is to examine only one of science's dimensions, and answers to these questions—even the most perfect of answers—do not necessarily constitute answers to the important problems raised by other dimensions of science. We may be of the opinion that answers to the epistemological questions are always going to be pertinent to these other problems; even more strongly, it may be held that a good epistemology of science is necessary, but not sufficient, to any successful understanding of the other aspects of science. This, however, is an open question. If our chief concern is the scientific institution's frightful complicity in what Michel Serres has called the "thanatocracy," then it is simply not obvious that we need to concern ourselves at length over the truthfulness or validity of this or that piece of scientific doctrine.[22] It should be obvious that in this case we *do* need to take very seriously science's ability to produce work that can be put to use in an instrumentally effective, indeed, a devastating, manner. Yet if our real concern is with the social facts and political realities, what we need to know is, for example, how it is that some thirty percent or more of scientists and engineers in most industrial nations today are working on weapons-related research; we need to know what the consequences of this state of affairs are likely to be, and how the dreaded outcome can possibly be avoided. A brilliant theory of demarcation in which the crucial criteria are at last formalized simply may not be of any great pertinence to such issues. And this is no doubt part of the reason why the former belief in the absolute centrality of the demarcation issue has been replaced in many circles by exactly the opposite opinion, so that the political and social conditions of science are thought to be the heart of the matter. This basic shift in attitude, by means of which the former centrality of epistemology has been rejected, is the key to what is referred to as the sociological turn in the philosophy of science.[23] An indication of this trend is the fact

[22]Michel Serres, "Thanatocratie," *Critique*, 298 (1972), 199–227.

[23]Actually, the expression may be overly restricted, for the trend in question is also a matter of a psychological, historical, and political turn. For an example of an extensive critique of the classical deductive-nomological model on largely psychological grounds (amounting to a rejection of the idealized image of the reasoning scientist), see David Faust, *The Limits of Scientific Reasoning* (Minneapolis: University of Minnesota Press, 1984).

that many researchers have conducted lengthy studies of science which at no point refer to the truth or falsehood of the knowledge in question, which is still called "knowledge" only out of a quasi-ethnographic respect for linguistic usage.[24]

An example of such a study is Richard Whitley's *Intellectual and Social Organization of the Sciences*, a work that describes the varieties of scientific "knowledge production" wholly in terms of the different ways scientific labor is organized, evaluated, and rewarded. From Whitley's perspective, the early, "intellectualist" views of science have long been surpassed, and it is no longer plausible to assume that science is unitary, that its results are justified in terms of logic, and that these results are progressive and cumulative. Even Thomas Kuhn's challenge to the epistemic logic of demarcation is now seen as having been overly limited, for if Kuhn's famous and highly influential work of 1962 pointed the way to a sociological analysis of the modes of knowledge production, many of its insights and concepts—such as that of "paradigm"—were still far too epistemic and intellectualist. Kuhn, for example, still privileged physics in the manner of the positivist philosophers of science, and he offered no accounts of how different sorts of knowledges are produced as a result of different social contexts and modes of organization. Kuhn's sociology of the scientific community is said to be particularly weak, for it rests upon an unrealistic notion of a homogeneous and coherent group of cooperative researchers who, in spite of their potential differences in worldviews, are all engaged in essentially the same process. According to Whitley, traces of these Kuhnian weaknesses still flaw some of the even more radical approaches involved in the sociological turn, which includes the work of the constructivists and the so-called strong programme of the Edinburgh school, animated by David Bloor. Although these tendencies seek to go beyond the epistemological perspective and to reveal the socially conditioned nature of scientists' judgments, Whitley holds that they have not yet achieved a convincing and detailed comparative sociology of scientific work. Whitley's conclusion, then, is that we will go on missing the sociological turn until the organizational structures of scientific labor have been delineated in detail; Whitley complains, for example, that in the constructivist work of Latour and Woolgar "the construction

[24]A work that prefigures the recent antiepistemological tendency in the sociology of knowledge is Florian Znaniecki, *The Social Role of the Man of Knowledge* (New York: Harper & Row, 1968), which establishes from the outset the sociologist's indifference to the truth or falsehood of the "knowledges" being studied.

of cognitive order is seen as an almost miraculous activity which defies further analysis."[25]

Such debates seem utterly remote from the former discussions of science in which the main argument was over whether justification was achieved by verification, confirmation, or only falsification. These two kinds of debate are indeed remote from each other, so much so that the question of their possible relation is a pressing problem. It should be easy to recognize that the existence of such different approaches to science is in fact a valuable thing; the emergence and development of sociological research into science is an important achievement, for it is a matter of investigating realities that were wrongly neglected as a result of the positivist and logicist biases. After all, many basic dogmas about science are seriously challenged once we focus on its social conditions; for example, a supposed value freedom or neutrality is revealed to be a kind of value investment and interestedness requiring examination. What seems erroneous, however, is the assumption that the new approaches to science can or should simply replace the former interest in the problems of epistemology.

It is crucial to distinguish here between two separate attitudes. The first is one in which the sociological turn is held to be worthwhile insofar as it seeks answers to very different questions than those taken up in an epistemological philosophy of science; at the same time, however, it is recognized that the sociological approach is not fully sufficient, and that the epistemological questions retain their pertinence. The second attitude, on the other hand, holds that the relation between these two approaches is much more conflictual; if the sociological turn does not answer the problem of demarcation at all, this is so because it shows that it is futile to look for the kinds of answers the philosophers want. Rather, the sociological turn leads us beyond such pseudoproblems by showing the question of truth to be of no importance to the realities of science; the sociological approach is fully sufficient, and the epistemological questions are ruled out.

For reasons that soon become apparent, I believe that the second of these two attitudes is incoherent and should be rejected, while the

[25]Richard Whitley, *Intellectual and Social Organization of the Sciences* (Oxford: Clarendon, 1984), 5. I highly recommend Whitley's book to literary critics, who may be interested to learn that they are working within organizational structures described as "fragmented adhocracies." For a clear and amusing presentation of the kinds of arguments at stake in what follows, see Michael Mulkay, "The Scientist Talks Back: A One-Act Play, with a Moral, about Replication in Science and Reflexivity in Sociology," *Social Studies of Science*, 14 (1984), 265–83.

first is correct and needs to be developed and elaborated upon. One may begin to expand upon this idea by pointing to a position that these attitudes have in common—a stance that distinguishes them from the logical positivist and logical empiricist conceptions of how to do philosophy of science. Both of the sociological attitudes I have in mind put in question the distinction that was formerly drawn between topics situated within the "context of discovery" and topics having to do with the "context of justification." This distinction, which was first formulated in these terms by Hans Reichenbach, is crucial to an understanding of the philosophy of science, for it is this distinction that underwrote the basic autonomy of the deductive-logical deliberations that long occupied the forefront in this field.[26] It was this distinction that set to one side the many troublesome problems related to the role of imagination and creativity within scientific discovery, as well as the massive difficulties having to do with the scientific division of labor, the institutional sites of research, and the consistent and intimate relation of that research to certain types of practical interests and nondeductive forms of inference. It was this distinction that established, on the other side, a wholly separate and unrelated problem, that of the logical reconstruction and justification of the isolated products of the other context, products usually referred to as "scientific explanations" and thought to be in need of translation into the form of deductive arguments.

If there has been a single predominant trend in this complex and fragmented field of inquiry over the past two decades, it concerns the idea that this basic dichotomy does not go without saying and can no longer be adopted in an unexamined manner. It is this dichotomy that is challenged by a wide variety of critics of logical positivism, ranging from Stephen Toulmin, Thomas Kuhn, and Jürgen Habermas to Isabelle Stengers, Karin Knorr-Cetina, and Frederick Suppe. Again and again, the basic point is made that there is no separate, abstract, epistemological context of justification, for both discovery and justification occur, insofar as they occur at all, in the institutional, historical, and sociopolitical context of the world within which science exists.

What becomes, then, of the questions that formerly animated the philosophy of science, questions having to do with the logical form of scientific explanations, with the nature and justification of scientific

[26]Hans Reichenbach, *Experience and Prediction* (Chicago: University of Chicago Press, 1938), 7. See also Harvey Siegel, "Justification, Discovery and the Naturalizing of Epistemology," *Philosophy of Science*, 47 (1980), 297–321.

knowledge, and with its relation to other kinds of discourses and knowledge claims? What becomes of what Popper called "Kant's problem," the problem of asking whether the limit can be drawn between dogmatic assertions and intersubjectively binding and valid knowledge of empirical matters?[27] It is in regard to these issues that the two attitudes I distinguished above diverge. For the one attitude seems to hold that a uniquely sociological account is all that is needed, and that no further epistemological questions about truth and justification are of any importance in regard to scientific work. Once we have an adequate social and psychological analysis of the social systems called science, we will be in a position to understand what motivates the scientists' various judgments and decisions—and this will be possible in the total absence of any arcane logical transcription of the propositions or statements supposedly implicit in the decision that this or that finding was ready for publication (or the decision that a published finding is correct or justified). The supposed *individual logic* of rational scientific decision making is replaced by a competitive, social process of "negotiation" and "construction"—a process that, unlike the reconstructive "logic of explanation," is local rather than global, contingent rather than necessary, biased and secretive rather than neutral and explicit, mechanistic and uncontrolled rather than rational and lucidly teleological.[28] It follows as well that if there is any line of demarcation between science and nonscience, the criterion is purely social and has nothing to do with the superior rationality or empirical soundness of the scientist's conceptual productions. Demarcation there might be, then, for the *institutions* of science clearly have their specificity, but such a demarcation is of a nonepistemological sort.

The Return of Epistemology

Are the epistemological questions having to do with the reference of knowledge claims to reality, with the truth and justification of

[27]Karl R. Popper, *The Logic of Scientific Discovery* (New York: Basic Books, 1959), 34.

[28]For examples of this sort of analysis, see Pierre Bourdieu, "The Specificity of the Scientific Field and the Social Conditions of the Progress of Reason," *Social Science Information*, 14 (1975), 19–47; Karin D. Knorr-Cetina, *The Manufacture of Knowledge: An Essay on the Constructivist and Contextual Nature of Science* (Oxford: Pergamon, 1981); David Bloor, *Knowledge and Social Imagery* (London: Routledge & Kegan Paul, 1976); and the many references provided in Whitley, *Intellectual and Social Organization of the Sciences*. One may also think of Michel Foucault in this context; for comments, see my n. 2, p. 150.

different beliefs and belief systems, finally eliminated or surpassed by the so-called sociological turn? I think not, and I suggest that it is possible to hold to the importance of the sociology of knowledge without assuming that it is right to brush aside these problems of classical epistemology. No one should advocate a return to a belief in the adequacy of the deductive-nomological model alone, but it is also crucial to avoid the weaknesses of the sociologism that has just been characterized, weaknesses that involve the return, within that program, of the very epistemological difficulties that were supposedly eliminated. These difficulties are twofold, for they concern at once the object domain of the sociology of knowledge as well as that discourse's own ambivalent status within that domain. I begin with this latter, more obvious issue, which can be introduced by making reference to some fairly recent methodological discussions within the sociology of knowledge.

According to the members of the Edinburgh school of the sociology of knowledge, a longstanding and constitutive shortcoming of that field was the idea that the sociological conditions of cognition need be identified only in regard to cases of failed or faulty cognition. As the story went, given some fantastic and obviously untrue mythical or ideological belief, it was of course necessary to inquire into the social and psychological conditions that could lead people to believe in such fictions. Yet in the cases of science's accurate and lawful understandings of natural phenomena, it was clearly reasoning and evidence about the phenomena themselves that dictated the explanations, and no social or psychological description of the discovery was needed. The innovation of the Edinburgh school's strong programme is precisely a rejection of this so-called arationality principle, which says that the sociological explanation only steps in where the "evidential rationality" of cognition fails.[29] According to David Bloor, the social causes are always present and are always determinant in the constitution of knowledge, even in those cases where the belief in question is thought to be true.[30] Consequently, Bloor imposes four requirements on an adequate sociology of knowledge: (1) it must provide causal explanations of the genesis of the beliefs it studies; (2) it must be impartial in evaluating these beliefs; (3) it must

[29]The "arationality principle" has been defended by Larry Laudan in his *Progress and Its Problems* (London: Routledge & Kegan Paul, 1977); for a recent criticism of his arguments, see Richard C. Jennings, "Truth, Rationality, and the Sociology of Science," *British Journal for the Philosophy of Science*, 35 (1984), 201–11.

[30]Bloor, *Knowledge and Social Imagery*, 14; see also James Robert Brown, ed., *Scientific Rationality: The Sociological Turn* (Dordrecht: Reidel, 1984), 9.

be symmetrical, in the sense of offering the same kinds of explanations for both true and false beliefs; and (4) it must be reflexive, that is, it must explain its own causal conditions.

In these terms, a first problem with the antiepistemological approach to the sociology of science has to do with the fourth requirement. This reflexivity requirement brings to the foreground the problem of explaining the emergence, among the various types of knowledge that can be observed, of that peculiar type of knowledge that purports to be a knowledge *of* knowledge, that is, a knowledge capable of *correctly* assigning all other knowledge claims to their place on the sociological map. Bloor's fourth clause requires that this discourse give a historical and causal explanation of its own emergence. Although this requirement indeed seems necessary, the problem of reflexivity does not stop there; rather, it leads the strong programme over into the very epistemological ground that sociological perspectives on science have sought to avoid—and to the question of the validity or truthfulness of the sociological explanations being proposed. The sociologist of knowledge must be prepared to respond to the inevitable question: Given your account of how everyone else's knowledge claims are in fact strategic or functional moves within a social arena, moves determined by social causes, why should we believe that your discourse is correct? By what miracle—or as a result of what interesting social and psychological conditions—does yours free itself from its social limitations and interests, thereby achieving objective validity? There is something very puzzling about assertive sociologies of knowledge that do not confront this question directly; as Peter L. Berger and Thomas Luckmann comment, it is like "trying to push a bus in which one is riding."[31] Here we find a perfect metaphor for the ambivalence of the sociology of knowledge, for its discourse situates itself implicitly within its own object domain, while it at the same time speaks of other discourses as if surveying them from outside the space or domain within which they are located (it is at once on and off the bus). If such a discourse has any pretensions whatsover to being realistic, then the domain within which it situates other discourses must be presented as *the* domain, that is, as reality; as such, this domain is also the only space within which the sociological account itself can exist, and hence it exists alongside or among the objects it explains "as if from above." If it claims that the objects and states of affairs within this domain are always causally

[31]Peter L. Berger and Thomas Luckmann, *The Social Construction of Reality: A Treatise in the Sociology of Knowledge* (Harmondsworth: Penquin, 1967), 25.

conditioned, it must recognize that it too is subject to such causal determinations. Consequently, what must be explained is its own difference; and most important, what must be presented is the basis or reason for the validity of its claim to be able to see what the other discourses do not see, namely, the causal conditions of knowledge. To show that this comment on Bloor's four requirements is not merely an abstract matter, we need only remark that we have a right to know how it was established, for example, that real explanations in sociology must be "causal." In what sense is this word used here? Although we may very well agree with this insistence upon a causal analysis, the idea that sociological explanations are always causal does not go without saying and is contested by any number of intelligent methodologists.[32] Here we see that the reflexivity requirement has epistemological implications that cannot be avoided by a consequential sociology of knowledge. To say this does not return us to the positivist position, for although the issues stressed here are clearly epistemological, it is possible to address them without assuming the existence of some purely autonomous context of justification. In other words, the problems of justification and of demarcation persist, even when the so-called context of justification is wholly inscribed, along with the context of discovery, within the sociohistorical world.

But this is not all, for the return of epistemological problems within the sociology of science does not have to do only with these "meta-" or reflexive issues concerning that discipline's methodology and justification. There are other epistemological problems that remain wholly "immanent" within the object domain of a sociology of science, problems having to do with the difficulty of providing a full explanation of the behavior of the object, science, within that domain. An important part of that behavior is science's ability to provide instrumentally reliable knowledge, and a full explanation of science must as a result explain this reliability.[33] In particular cases where this reliability breaks down, it is also necessary to explain why this occurs. We are now discussing Bloor's third requirement, which demands a symmetrical explanation of true and false beliefs. This

[32]Compare, for example, Mario Bunge's contention that causality is a variety of determination that is not frequently appropriate within the social-historical domain: *Causality and Modern Science* (New York: Dover, 1979), 262–76.

[33]This general realist strategy has been developed in great detail by Richard N. Boyd; see his "On the Current Status of the Issue of Scientific Realism," *Erkenntnis*, 19 (1983), 45–90; and the articles cited in the next chapter. I recommend Boyd's work to literary scholars interested in the philosophy of science, for they will find here a rigorously argued and well-informed alternative to the false choice between positivism and constructivism.

requirement is no doubt intended to fill the lacuna perceived in earlier sociologies of knowledge, that lacuna left by the aforementioned tendency to seek social causes only for false beliefs. Yet we must also ask whether there is a further, perhaps unintended implication in this clause, for we wonder whether it is meant to suggest that sociological explanations of true and false beliefs are essentially the same in some deeper sense. In other words, this clause seems useful insofar as it points to an important and neglected area of inquiry, yet the notion of "symmetry" could serve to close off another important issue, namely, that of the *differences* between true and false, or at the very least, successful and unsuccessful, explanations. These differences need not be purely epistemological, and it would certainly be interesting to find out that the social conditions of valid discoveries were consistently different from those behind bad science, mythologies, rank ideology, and so on. Bloor's second clause, which requires that the sociologist somehow be impartial, could be read as suggesting that the question of the difference between true and false explanations is not a topic falling within the sociology of science, which would simply bracket this topic and go on with its search for causal explanations of all knowledge claims. This, we are arguing, would be a mistake, for the difference between good and bad science—or more generally, *the possible existence of an effective epistemic demarcation* within the field of knowledge claims—is also something that an adequate sociology of knowledge should attempt to investigate. Bloor's third clause, then, becomes a much more complicated affair when revised in this light and greatly adds to the sociologist's tasks. Before I elaborate on that, however, I must fill in some steps in the above argument with somewhat greater detail. What is meant, for example, by "instrumental reliability", and why does that seemingly pragmatic matter necessarily involve epistemological issues? An example may help to clarify this point.

In aeronautical engineering it is useful to conduct tests of new aircraft designs by using scale models in a wind tunnel instead of flying a full-sized prototype, which would be costly and dangerous.[34] For such a process of substitution to be effective, it is important to know whether the results of the test can accurately represent the pertinent features of the prototype's future performance. Here we have a splendid example of the problem of representation that is so

[34]For background, see the articles "Reynolds Criterion," "Airfoil," "Wind Tunnel," "Aerodynamics," in *Van Nostrand's Scientific Encyclopedia* (Princeton, N.J.: D. Van Nostrand, 1968).

central to literary theorists; the little scale model surely *is not* the huge airplane—which does not even exist yet—and can hardly manifest its "full presence," yet the engineers seem to think that, if they conduct certain wind tunnel tests on the model, they will find out something about the flow of the atmosphere around the body of an aircraft, about the lift provided by its wings, about forces such as drag to which it will be subjected, and all of this relative to different speeds and atmospheric conditions (such as density and temperature). Such wind tunnel testing is no simple matter, and it is not by means of some kind of everyday perception that scientists and engineers make the observations that they will then extrapolate to the scale of a real airplane. Only an expert knows what to look for and what the measurements signify, and the expert knows this only because a large body of theory is at work in making these observations coherent and meaningful. These theories have to do first of all with whether the model situation is indeed sufficiently similar to the real conditions to provide the basis for any reliable extrapolations or predictions. For example, only if certain conditions are fulfilled do the measurements of the forces of lift and drag on the model serve as an adequate basis for calculating coefficients that accurately apply to the full-scale prototype, and these conditions are specified in terms of complex aerodynamic theories. These latter involve, for example, the theory of fluid dynamics, with its distinction between laminar and turbulent flows, and the ratio (known as the Reynolds number) describing the conditions under which there is a transition between them. It turns out, for example, that it may be necessary to use very high speeds in the wind tunnel in order to obtain measurements having an adequate similitude to the full-scale prototype.

Here we have an example of an experimental practice that, although it relies upon a large body of aerodynamic and, more generally, physical theory, displays a high degree of instrumental reliability. Bluntly put, the extraordinary achievements of the science and practice of aviation, a twentieth-century phenomenon, have been greatly facilitated by the consistent efficacy of this kind of modeling—even though no one assumes that any model can offer a perfect representation of a real airplane's performance. And aviation as a whole is an excellent example of the massive presence within our everyday world of inventions that make manifest the instrumental reliability of specific scientific theories; only the radical skeptic can doubt that airplanes built in function of these theories do in fact fly, and that even in cases of failure it is the machines (and human beings) who are at fault, not the basic aerodynamic principles. It is likely, for example, that a large percentage of the readers of this book will have at some point relied upon this "useful" product of scientific research and experimentation. What is interesting, then, is not the question of the existence of this instrumental reliability—which only the wild skeptic can seriously contest—but the problem of its explanation. It is precisely this problem that leads the sociologist of science directly to the kind of epistemological issue that was evacuated with the blunt equating of knowledge and cognition or belief.

Let us suppose that the sociologist's questions in regard to this example can be formulated as follows:

> Q1: Why has modern society developed a science of aerodynamics, and why does this body of knowledge consist of the specific set of beliefs, propositions, and theories that we associate with that name? In other words, why is there an aerodynamic doctrine at all, and why this one?

What I want to contend is that a full answer to these questions requires an answer to the following kind of question as well:

> Q2: What makes this aerodynamic doctrine have the instrumental reliability that it has demonstrated? Why do things like airplanes generally work when we base their design on this doctrine?

I do not think that an answer to Q2 itself constitutes a sufficient answer to Q1, and herein resides an irreducible difference from the positivist and logical empiricist visions of science. Rather, the theory or beliefs that amount to an adequate answer to Q2 are a proper subset of those that respond to Q1. For example, it could be that the prevalence of the social activities we call commerce and warfare had a lot to do with bringing about the science and practice of aviation, because flying machines quickly revealed themselves to be useful to both. Thus a detailed social analysis could show how very specific kinds of interests led to a channeling of research in certain directions, to the exclusion of other possible paths of inquiry and other investments—be they scientific or non-scientific. The real issue, then, is whether an answer to Q1 is already an answer to Q2, or more generally, whether there is another kind of issue to be associated with Q2. I believe that there is. It seems that if we say that a detailed social analysis provided in response to Q1 also answers Q2, additional assumptions must be brought in, because Q2 does not at first glance pertain only to states of affairs that are immanent to the social level of description. Rather, it seems that Q2 concerns the relationship between a set of beliefs and the natural conditions to which these beliefs refer—be this reference relation partial or complete, imaginary or real. If we are to eliminate our epistemological subset, or Q2, subsuming it entirely beneath the sociological account given in response to Q1, we need to assume in advance that the instrumental reliability of aerodynamics is simply illusory, or that it can be explained without in any way granting that this doctrine owes its reliability to its reference to natural (or nonsocial) realities. But this assumption that would preclude the epistemological question of the scientific doctrine's *reference* to nature in fact amounts to an extremely dogmatic assertion of *one* answer to that very question. We are asked to believe that the pioneers of aerodynamics were mistaken, that they were wrong in thinking that their knowledge was about an external nature; in spite of this, inventions based on these theories worked, and the airplanes constructed with these false beliefs in mind still flew. It is not at all clear why we should prefer that dogmatic assertion to an exploration of the question it precludes, especially given the implausible nature of the conception it provides for us. Are we to imagine that airplanes can fly only because they are useful in warfare and business? A perpetual motion machine would also be very useful in war and business, but the various attempts to construct one have all failed. Are we asked, then, to imagine that we believe in the flight of aircraft because the theories upon which they are based "overdetermine the observations?" But a lot of people who fly around in airplanes and believe that they see them flying know nothing at all about aerodynamic and physical theory, they could not recognize Bernoulli's equation in a crowd, so how is it that they are seeing the world through that theory? Perhaps what should be imagined is that we need not bother with the supposed reference of aerodynamic theory to nature because the people who got the first airplanes off the ground did not rely upon such abstract theoretical fictions as forces and instantaneous velocity. Or did they? It seems more likely that they did find some of the basic notions of fluid dynamics to be useful; it even appears that the principles that explain how a wing provides lift when moved through the air at certain speeds had a role in the design of the earliest airplane wings. It is doubtful that the earliest pilots and aerodynamic theorists thought that lift was only a social fact; we should note as well that the theory of lift was quantified a year before the first successful flights in 1903.

It seems that, for our example, the attempt to close off the question of scientific realism by means of a sweeping assertion is bankrupt. It is simply cavalier and foolish to assume that the reality of flight is somehow reducible to the causal conditions of culture, society, psychology, or belief, even if without these "determinants" airplanes would never have been invented and the only flight would have been that of birds, bats, clouds, moribund leaves, insects, pollen and seeds, and certain kinds of squirrels and fishes. A comprehensive explanation of the historical emergence of modern aviation and of the scientific theories involved cannot simply evacuate the question of the referential status of the doctrine in question. On the contrary, it seems that our aerodynamic theories have to do with the relationship between certain human inventions, on the one hand, and an objective environment—the earth and the sky, if you like—on the other.

Although the case of aviation and the related scientific theories is only one example, it is hard to imagine an adequate account of the natural sciences that would leave it out. Thus I feel warranted in saying that this example provides an adequate basis for considering the strengths and weaknesses of a perspective on science that excludes an epistemological approach, and with it, the complex and difficult issues surrounding the nature and status of a scientific doctrine's reference to an independent natural reality. The full-blown sociological and constructivist account of science cannot offer a plausible explanation of an instrumental reliability that it necessarily acknowledges as belonging to what must be explained; consequently, this type of account—no matter how valuable it may be in filling in a previous blindspot—is incomplete. Its rejection of epistemology amounts to a dogmatic assertion that there can be no justified knowledge because no knowledge can be linked to any evidence other than that of its own framework.[35] Here we see that the account under consideration is another version of framework relativism. Thus, if our earlier arguments against that position were correct, we can conclude that this whole conception of science is not only incomplete but incoherent as well. Furthermore, the strong programme is often presented as evidence of the validity of framework relativism—which would then be the expression of the best account

[35]Thus Bloor, in his *Wittgenstein: A Social Theory of Knowledge* (London: Macmillan, 1983), praises Durkheim for having pointed to the social bases of categories of thought and systems of belief but criticizes him for having held that under some social conditions the truth of a belief can be a condition of its social acceptance. There is no window on the world, and whatever world there is is the one constructed by the play of social interests (3, 156). Bloor states: "Objectivity and rationality must be things that we forge for ourselves as we construct a form of collective life. So the work of Copernicus is undone. Human beings are back in the center of the picture" (3).

of the current scientific world view, one in which science's previous claims to know nature objectively would have been abandoned. Yet my demonstration of the inadequacy of the strong programme's account of science entails that this program can offer no independent support to framework relativism.

In short, anyone seriously interested in exploring the models of knowledge presented to us by the natural sciences must confront the question of the relationship between these sciences and the realities to which they are meant to refer. The properly epistemological issues concerning the nature of scientific explanations, their truth and reference, cannot be avoided, and it is to them that I turn in the next chapter.

CHAPTER 3

On Scientific Realism

As the following discussion of natural scientific knowledge obviously can provide only a limited account of such a vast subject, its scope is purely tactical and metacritical.[1] My aim, then, is to contest the supremacy of a conception of science that is often evoked by literary scholars as if it somehow had already been established as binding and authoritative. The conception I wish to contest is frequently thought to lend support to framework relativism, and this same conception of science is often linked to the reductive sociological approaches criticized in Chapter 2. Humanists who favor this conception of science typically believe that it is the only appropriate rebuttal to the positivist myths surrounding natural science—myths, for example, of cumulative and lasting results, and of progress and objectivity. It is also thought that, as a consequence of this new conception of the natural sciences, the humanities and the sciences are once more set on an equal footing.[2]

[1] A useful bibliographical tool is Richard J. Blackwell, *A Bibliography of the Philosophy of Science, 1945–1981* (Westport, Conn.: Greenwood, 1983). See also the excellent bibliographies published annually by *Isis*.

[2] Two examples: In his bold essay "Science moderne et interrogation philosophique," Cornelius Castoriadis begins by suggesting that, given the fundamental conceptual crises that characterize the modern natural sciences (basically, Gödel, quantum mechanics, and Einstein), it is foolish to think that the sciences of man have any catching up to do; in *Les carrefours du labyrinthe* (Paris: Seuil, 1978), 147–217. A more recent declaration on the science–humanities relation is even more blunt: "The recent epistemology of the *Naturwissenschaften* (I am thinking of Kuhn, of course, Feyerabend, Lakatos, etc.) points towards a dissolution of this distinction"; Gianni Vattimo, "Myth and the Destiny of Secularization," *Social Research*, 52 (1985), 347–62, citation: 357.

Some of the major weaknesses of framework relativism and sociological reductionism have been brought forth in Chapter 2; the essential epistemological position often associated with them must now be brought into direct confrontation with the realist alternative.[3] This antirealist epistemology is referred to in what follows as "constructivism." Many of the authors who defend the views I have in mind may not employ this particular label, but the theses on science they defend are aptly placed beneath this rubric.[4] What, then, are the main theses of the constructivist position to be criticized? Two essential tenets can be identified at the outset. The first, a rejection of a basic realist notion, amounts to saying that the natural sciences have not in fact achieved any true knowledge of a mind-independent natural reality. On the contrary, science's methods are said to be wholly relative to a theoretical framework, and thus science amounts to a construction, not a discovery, of realities. Second, the constructivist insists upon the heterogeneity of the theoretical frameworks that determine science's findings. As a result of this heterogeneity, it is supposed to be impossible to argue for the commensurability of the various results obtained by different scientific theories, which turn makes it impossible to argue for the continuity required by the progress of knowledge. It follows from the extreme versions of constructivism, finally, that the various scientific doctrines collectively enjoy no special epistemic status in regard to a framework-independent "nature"—and that there can be no successful epistemological demarcational arguments.

My goal in this chapter is to challenge literary scholars' reliance upon these constructivist tenets. For the purposes of this challenge, I need only establish a rather minimal claim, namely, that it is not

[3]I am especially indebted in what follows to the work of Richard N. Boyd, whose presentation of the issues is the clearest I have found, and whose arguments for realism are detailed and, in my view, correct. See his "Metaphor and Theory Change," in *Metaphor and Thought*, ed. Andrew Ortony (Cambridge: Cambridge University Press, 1979), 356–408; "Realism, Underdetermination, and a Causal Theory of Evidence," *Noûs*, 7 (1973), 1–12; "Lex Orandi Est Lex Credendi," in *Images of Science*, ed. Paul Churchland (Chicago: University of Chicago Press, 1985), 3–34; "The Logician's Dilemma: Deductive Logic, Inductive Inference and Logical Empiricism," *Erkenntnis*, 22 (1985), 197–252; "Observations, Explanatory Power, Simplicity," in *Observation, Experiment, Hypothesis in Physical Science*, ed. Peter Achinstein (Cambridge, Mass.: M.I.T. Press, 1985), 45–94; and especially, "On the Current Status of the Issue of Scientific Realism," *Erkenntnis*, 19 (1983), 45–90. For a well-informed argument in favor of a position that seems wholly consonant, see David Gooding, "How Do Scientists Reach Agreement about Novel Observations?" in *Studies in the History and Philosophy of Science*, 17 (1986), 205–30.

[4]For a sample of constructivist epistemology explicitly presented as such, see Paul Watzlawick, ed., *The Invented Reality* (New York: Norton, 1985).

reasonable to conclude that constructivism presents a correct and adequate account of the relation between science and nature. To make good on this minimal claim, it is not necessary to prove conclusively that we are already in possession of the definitive response to the question of the relation between science and nature. I do not have to prove that some scientific doctrine represents the ultimate truth about some aspect of nature, nor do I need to set up the complete framework for a comprehensive philosophy of science. Rather, the minimal claim only aims at assaulting the certainty and dogmatism of the opposite viewpoint, following which the truth of the natural sciences should be univocally interpreted as a *construction*. There are two basic ways such a minimal claim can be supported: negatively, by examining the weaknesses in the rival account, and positively, by establishing the plausibility of alternative conceptions. Hence, in addition to arguing for the minimal claim, I at the same time present arguments for a maximal claim. What is involved in the latter are different reasons for which some version of a realist account of science should be preferred. I do not pretend to be able to provide an irrefutable proof of this latter claim, but I am convinced that the arguments supporting it are much stronger than those available to the other side of the question. The conclusion that should emerge from this discussion is that arguments for framework relativism in relation to the human sciences do not find any real support in the contention that such a relativism also holds in the natural sciences—for no such thing has ever been properly established. It will not do, for example, to say that, because it has been shown in the natural sciences that all observations are theory-laden, it follows that all textual evidence in literary and historical debates is totally overdetermined by the reader's perspective, theory, or prejudices. This is sloppy thinking, first because the real meaning and validity of the antecedent clause remain to be established, and second because the analogy in question obscures too many crucial considerations to be very instructive about the problems of literary knowledge. Given that the argument just mentioned is in fact a commonplace of literary theory, the following discussion of the philosophy of science is in fact central to the state of the art in the literary disciplines.

I begin my discussion by introducing the topic of realist philosophies of science. Some of the different ways of construing this position are presented against a broad background of issues, before a specific version of realism, one defended by Richard Boyd and others, is isolated. Only then do I present some of the central epistemological arguments for and against realism. The chapter concludes with a further discussion of what is *not* entailed by a defense of the special status of science's knowledge claims.

Variations on Realism

As literary critics know, "realism" is in general a vague and overworked term, and the situation within the philosophy of science is perhaps no different, in spite of various thinkers' efforts to stipulate the essential attitudes and theses that can rightly be associated with the word.[5] Nonetheless, I do not think that the term should be abandoned, for it is simply not clear what other label could be used to evoke the basic stances associated with it. In spite of the significant differences in detail, there are some worthwhile points of convergence in certain of the recent specifications of the central realist tenets. The two main principles of scientific realism have been recapitulated in the following manner: (1) the terms of mature scientific theories typically refer; and (2) the laws of mature scientific theories are typically approximately true.[6] As other authors have been quick to point out, such statements of basic tenets are deceptively simple, for they are more accurately seen as formulas containing free variables (such as "laws," "refer," "true") that, when given specific arguments, yield a variety of outcomes. For this reason it might be more accurate to follow Peter Smith in identifying the first two basic tenets of realism as problems or questions: (1) Do scientific theories describe things that exist? and (2) Do scientific theories have truth and falsehood? Smith labels these problems as those of "reference" and "predication," respectively, and adds to them a third, that of "progress."[7] Thus the issues surrounding realism have both synchronic and diachronic dimensions, and the notion of progress is

[5]In addition to the Boyd articles cited above, see Norman Stockman, *Anti-Positivist Theories of the Sciences: Critical Rationalism, Critical Theory, and Scientific Realism* (Dordrecht: Reidel, 1983), 72–94; and Hilary Putnam, "Three Kinds of Scientific Realism," *Philosophical Quarterly*, 32 (1982), 194–200. One may also consult Jerrold L. Aronson, *A Realist Philosophy of Science* (New York: St. Martin's, 1984), where the problem of confirmation is foregrounded and a "comparative maximum likelihood" approach is developed. For a broad overview, see Ian Hacking, *Representing and Intervening: Introductory Topics in the Philosophy of Science* (Cambridge: Cambridge University Press, 1983).

[6]Attributed to Richard Boyd by Hilary Putnam in *Meaning and the Moral Sciences* (London: Routledge & Kegan Paul, 1978), 20–21. For references to Boyd's own published statements of his views, see note 3 above.

[7]Peter Smith, *Realism and the Progress of Science* (Cambridge: Cambridge University Press, 1981), 1–4.

seen as having a direct bearing on the way we should understand science's reference as well as its truth. According to Smith, many realists seek to make their answers to the first question serve as the basis for their answers to the second and third questions. The legitimacy of such a procedure is contested by some opponents of realism, who counter by stressing the priority of the epistemological conditions; at the other extreme are realists who believe that reference is too epistemic a notion to capture the ontological thrust of realism, which concerns not what is known but what is (thus Geoffrey Hellman usefully distinguishes between pure and semantic types of ontological realist theses[8]). Such remarks assume that a realist position necessarily involves both an epistemological and an ontological component, for realism makes certain claims about what exists as well as claims about our ability to know what is the case. It should indeed be clear that all such matters are complicated by the basic difficulties attendant upon the distinction between an epistemology and an ontology, as well as by the absurdities that result when no such distinction is drawn. C. S. Peirce thought it was nonsensical to speak of a being existing outside of the totality of what we can experience or think about, yet only an idealist holds that the two orders are coextensive, or, for example, that there could *be* no nature in the absence of sentience (Peirce's "synechism," which posited a progressively more orderly and sentient nature, was indeed a form of idealism).

The relation between the ontological and epistemological facets of realism quickly leads to the thorniest problems, and the different strains of realism, as well as the spectrum of antirealist positions, should initially be introduced at this level of generality. Although many of the more technical discussions of realism within the philosophy of science seem to turn about the limited issue of the semantics of theoretical and observational terms, this issue is in part ontological. A failure to acknowledge this fact may be behind the way this discussion tends to bog down, and several authors have stressed the need to engage in a broader interrogation of the issues, one in which the more basic presuppositions are explored and assessed.[9] We cannot, then, begin with any one of the concise specifications of realism in which the opposition between theoretical and observational terms is taken for granted. Although such a procedure could help us to

[8]Geoffrey Hellman, "Realist Principles," *Philosophy of Science*, 50 (1983), 227–49.

[9]Extensive discussions of this issue are to be found in Frederick Suppe, ed., *The Structure of Scientific Theories* (Chicago: University of Illinois Press, 1977); and Raimo Tuomela, *Theoretical Concepts* (Vienna and New York: Springer, 1973).

arrive quickly at the debate between the realist, instrumentalist, empiricist, and constructivist positions, this might only be achieved at the cost of clarity. Instead, I first discuss some understandings of realism in which the position is yoked to the question of ontological commitments; this discussion should make it possible to isolate one, highly tenable form of realism from among the other varieties, at which point we can move on to the debate between realism and constructivism.

We may begin, following Raimo Tuomela and Wilfred Sellars, by making a broad, nontechnical distinction between the "order of being" and the "order of knowing," for such a distinction can be used to provide a characterization of some of the predominant positions on scientific knowledge.[10] It is in relation to these orders that the priorities of diverse positions can be charted. I begin with the rough dichotomy drawn by Sellars between what he calls the "manifest" and the "scientific" images of the world before moving on to distinguish between more nuanced positions. To get at this first dichotomy, we need only take up one of the *loci classici* in the field, namely, the discussion of Eddington's desk. Sir Arthur Eddington, the reader may recall, discussed the difference between two stories about a desk, the first one being the familiar story about an everyday piece of furniture, and the second story being one told by a scientist. In the first story the desk has color, weight, and solidity; in the second, the same desk reappears as "a host of tiny electric charges darting hither and thither with inconceivable velocity. Instead of being solid substance my desk is more like a swarm of gnats."[11] The question concerns the relation between these two desks. To return to the previous terminology, the first desk is that of the manifest image, and the second is the one identified as its latent reality following *one* understanding of scientific theories. As described by microphysical theories, the desk is not even as familiar as a swarm of gnats, for it would really be a cloud of elementary particles or quantons in a volume of four-dimensional space-time. *That* understanding of scientific theory is precisely what is defended by some realists and contested by other philosophers of science; thus Husserl speaks of the abstract and artificial *Ideenkleid* ("cloak of ideas") of physical theories in which the naturalist attitude veils the fullness that our

[10]Raimo Tuomela, *Science, Action, and Reality* (Dordrecht: Reidel, 1985); Wilfrid Sellars, *Science, Perception, and Reality* (London: Routledge & Kegan Paul, 1963); and *Science and Metaphysics* (New York: Humanities, 1968).

[11]Sir Arthur Eddington, *New Pathways in Science* (London: Cambridge University Press, 1935), 1.

experience has in the *Lebenswelt* ("life-world").[12] And at the other extreme, that of a reductionist materialism, Paul M. Churchland claims that common sense and everyday perceptions are in fact only the theories that "got there first" and should not be made the touchstones of all other conceptions.[13] For Churchland, excellence of theory is the yardstick of all ontology, and when our common and everyday observations diverge from the stipulations of the best available scientific theory, we should opt for the latter. Thus a loud noise is better understood as "large amplitude atmospheric compression waves," and to say that an object is red is a crude shorthand for the more accurate expression, which is that the object "selectively reflects EM waves at $.063 \times 10^{-6}$m".[14]

Thus one kind of extreme realist position holds that scientific theory should be granted priority over other conceptions both in regard to epistemology and ontology: what is known is known best through science, and what the best scientific theories refer to is all that there is. Empiricism offers the opposite conception, for it grants priority in both cases to the manifest, ordinary, or everyday image of the world. It specifies first that something like direct or immediate observation, and not conceptual extrapolation, is the basis of all our knowledge, and second that referents of any concept that cannot be rigorously linked to this kind of perception are not good candidates for ultimate reality; until they have been spotted, there is no reason to believe in gluons. Similarly, the instrumentalist position contends that theoretical terms should not be thought to refer to existing realities unless they can be linked to direct observation. Theories, then, are only instruments that help us to systematize empirical observation statements, and the realist errs in thinking that the entities referred to in scientific theories (such as invisible forces and particles) are things in themselves. We can easily illustrate the empiricist position by returning to our previous example of aviation. All that is indubitable, the empiricist would say, is the intersubjectively reiterated observation of the gross everyday reality of flight, but such

[12]Edmund Husserl, *The Crisis of European Sciences and Transcendental Philosophy: An Introduction to Phenomenological Philosophy*, trans. David Carr (Evanston, Ill.: Northwestern University Press, 1970), 51.

[13]Paul M. Churchland, *Scientific Realism and the Plasticity of Mind* (Cambridge: Cambridge University Press, 1979), 44.

[14]Churchland, *Scientific Realism*, 28–29; this work abounds with such examples. Instead of hearing the roar of the surf, people at the beach "listen to the aperiodic atmospheric compression waves produced as the coherent energy of the ocean waves is audibly redistributed in the chaotic turbulence of the shallows," and when the sun sets, they admire "the wavelength distribution of incoming solar radiation shift towards the longer wavelengths," 29–30.

observations are not sufficient evidence for making metaphysical assumptions about the existence of some of the entities, such as vortices and forces, postulated in aerodynamics.

It should be noted right away that, in many of the theory–observation debates in the philosophy of science, the two positions just described—roughly, instrumentalist-empiricism and extreme realism—are the two primary opponents. Yet several other positions are also commonly defended. A constructivist position, for example, resembles the realist position insofar as both deny the priority of ordinary observation or the manifest image in epistemology; the manifest image, it contends, is itself theory-laden, and all observation is made relatively to some conceptual system. Yet constructivism diverges from realism in other crucial respects. On matters ontological, it preaches abstinence (no knowledge constructs should be thought to refer to noumenal realities) or promiscuity (there are as many realities as there are constructs). Whence an exhortation in passing: do not forget that rejections of the possibility of a sharp dichotomy between theoretical and observational terms can be uttered in the name of very different positions and need not be meant to support either idealism or constructivism. Thus Raimo Tuomela, who characterizes himself as a realist, denies that we can engage in nonconceptual yet cognitive epistemic commerce with the world. There is no infallible form of immediate perception or direct observation, and attempts to establish a sharp theory–observation dichotomy ignore the context-dependence of the notion of "observability," which in science often involves special conditions and training having to do with complex instrumentation and an elaborate conceptual background knowledge. In this vein Tuomela goes on to define (following Sellars and Putnam) an "internal realism" that is said to be distinct from "metaphysical" realism as well as from empiricism and constructivism. This "internal" or "weak" realism contends that both descriptions of Eddington's desk refer, even if the scientific description is in some contexts more accurate and truthlike. Tuomela does not say so, but it strikes me that he should want to add the converse as well: in some contexts, it is the desk as described by our manifest image that is most accurate and precise—for these are epistemic notions that are relative to the question being posed. There is no such thing as absolute exactitude in the measure of nature, and its quest is an incoherent fantasy.[15] Thus, if science is in general

[15]Wittgenstein makes this point in various remarks, for example, *Zettel* (Berkeley: University of California Press, 1970), § 438–41; *Philosophical Investigations* (New

given epistemological priority (and indeed, Tuomela's regulative maxim is *scientia mensura*), this very epistemology need not specify a reductionist ontology in which one level of description is held to be ultimate or fundamental.[16] Although internal realism holds that scientific theories approximately denote real, mind-independent entities that are not part of the manifest image, this proposition does not entail a belief that only such entities exist. Tuomela specifies the ontological thesis of this "minimal scientific realism" as follows:

> (MSR) All sentient and non-sentient physical objects have at least the constituents and properties which correspond to the scientific terms needed in the theories which best explain the overall behavior of those objects.[17]

Given this position as a starting point, it is possible to delineate stronger forms of realism by means of two independent steps. First, the "internalism" can be replaced by a stronger version in which the realist adopts some version of a correspondence theory of truth (e.g., a causal theory of reference), following which it is incorrect to define truth uniquely as an epistemic relation (such as that of idealized justification). In my view, excellent arguments have been given for this move, arguments basically showing that the internal realist's opposition to so-called metaphysical realism itself rests upon properly metaphysical assumptions—beginning with the assumption of an unknowable noumenal reality.[18] Second, the minimal realism can be changed by limiting the ontological commitments of scientific theory to certain levels of description. The motive for moving in

---

York: Macmillan, 1968), §§ 88, 91; the observation is supported by the careful analysis of quantification and measurement in Marx W. Wartofsky, *Conceptual Foundations of Scientific Thought* (New York: Macmillan, 1968), chap. 7; the same theme is flamboyantly exploited throughout Michel Serres, *Le Passage du nord-ouest* (Paris: Minuit, 1980); in English, see Josué Harari and David Bell, eds. and trans., *Hermes* (Baltimore: Johns Hopkins University Press, 1982); and *The Parasite*, trans. Larry Schehr (Baltimore: Johns Hopkins University Press, 1982).

[16]For some background to this notion of levels of description, see Mario Bunge, "The Metaphysics, Epistemology and Methodology of Levels," in *Hierarchical Structures*, ed. Lancelot Law Whyte, Albert G. Wilson, and Donna Wilson (New York: American Elsevier, 1969), 17–28.

[17]Tuomela, *Science, Action, and Reality*, 126. Clearly a clarification of the notion of "explanation" is crucial in making sense of this slogan. For discussion, see Chapter 6.

[18]I do not rehearse these arguments in detail here, as they are not essential to the present context. I do, however, refer in what follows to the stronger, externalist forms of realism. See Ruth Garrett Millikan, "Metaphysical Anti-Realism?" *Mind*, 95 (1986), 417–31. For Tuomela's relativizing of truth to a conceptual schema, see "Truth and Best Explanation," *Erkenntnis*, 22 (1985), 271–99.

this direction is simple: some of the entities proliferating in the manifest image are illusory, and the minimal version does nothing to identify them. Tuomela goes on to define "moderate" and "extreme" scientific realisms as follows:

> (MOSR) (a) Minimal scientific realism is true, and (b) non-sentient physical objects have only the constituents and non-relational empirical properties which correspond to the theoretical scientific terms needed in the best-explaining theories.
>
> (ESR) All non-sentient physical objects and all sentient objects (including persons) are "mere" scientific objects.[19]

The moderate version is stronger than the minimal in that it excludes from the category of real, mind-independent existence the designata of such predicates as "owned by the government"—which does not refer to any particular material, ontologically independent entity. Such an ontological decision is based on the idea that some entities in the manifest image are in fact imaginary. That they are purely relational, representational, or fictive can be further specified by saying that they are not capable of engaging in causal interaction with other entities; thus, if "owned by the government" seems to refer to a real state of affairs, this is only because this description offers a global and relational reference to various other entities that do exist. This example, which is ruled out by both the moderate and extreme brands of realism, in fact straddles the borderline between them given that it involves a predicate that is relational and that has to do with some concept of collective human agency (the government). Although the moderate version may not rule such entities out (is the government a physical object? is it sentient?), the extreme version certainly does. In fact, it rejoins the position of Churchland mentioned above in that it rules out all predicates refering to such secondary qualities as colors, tastes, and sounds, which exist only relative to sentient organisms and should be "reduced" to lower-level physical, materialist explanations. Along with the language of so-called sense qualities, this extreme, reductionist ontology excludes all predicates having to do with intentionality, personhood, values, and action. Here we see the danger of yoking a realist conception of science to ontological commitments of this sort, and it is crucial to distinguish between realisms that take this step and those that do not. Literary scholars who burke this distinction in their haste to refute the hated realism make a basic error.

[19]Tuomela, *Science, Action, and Reality,* 126.

In the present context the point of sketching these ontological variations on the realist theme is not to try to solve the matter but to demonstrate the diversity of the conceptions involved. It is preemptive and limited simply to equate a realist philosophy of natural science with the most extreme of these views. Realism need not mean reductionist materialism, and to argue brilliantly against the latter is not necessarily to win points against the former. In what follows, I focus on a moderate and nonreductionist brand of realism, the main ontological tenet of which is that there is a mind-independent reality having entities that take part in causal interactions, and the central epistemological tenet of which is that these things are knowable, albeit partially and by successive approximations only. Scientific truth, on this view, is a matter of a relation of correspondence or reference between the theoretical and factual claims of science and these mind-independent realities. Finally, the progress of the advanced sciences is the result of successively more accurate approximations to the truth about both observable and unobservable phenomena. As Boyd states it, "later theories typically build upon the (observational and theoretical) knowledge embodied in previous theories."[20] I leave the other fascinating ontological issues open, although some of them concern me when I turn to the the status of some entities postulated in literary criticism.

I now present a brief and highly schematic survey of some important positions before moving on to some of the arguments between them:

Empiricism places epistemic and ontological priorities on the manifest image, or ordinary perception, and requires a sharp theory–observation dichotomy. Its theories are confirmed only by observations deductively predictable from the theory to be tested.

Reductionist materialism places epistemic and ontological priorities on a fundamental or ultimate scientific (physical) theory. It denies any sharp theory–observation dichotomy and requires intertheoretical continuity of reference. Reductionist materialism is only one variant of realism.

Realism places a degree of epistemic priority on scientific theory, but not exclusive of other warranted knowledge claims. It is nonreductionist and does not postulate a single, fundamental or ultimate level of description. Its ontology is multileveled: scientific theories refer to the real, but not perfectly, and not to the exclusion of all other (e.g., manifest) images. It denies any sharp theory–observation

[20]Boyd, "Current Status," 45.

dichotomy insofar as theoretical terms are held to refer to mind-independent realities, and its theories are interpreted realistically. Intertheoretic continuity of reference is argued to be the basis of progress of advanced science.

Constructivism places epistemic priority on a plurality of conceptual frameworks, which overdetermine methodology and evidence. It denies any sharp theory–observation dichotomy and continuity of reference between theories or frameworks. Its ontology is idealist-pluralist, or agnostic.

Theory Incommensurability, or Against Realism

It should be evident that the realist position we have in view has been criticized from a variety of other positions. In what follows I am primarily concerned with the objections that are thought to support a constructivist alternative to realism. This means that I am not overly concerned with the quarrel between empiricism and realism—not because this dispute is uninteresting or unimportant, but because it does not directly influence my central argument. It should nonetheless be pointed out in passing that it is probably the shortcomings of a logical empiricist approach to science which have set the stage for the constructivist and framework relativist tendencies. Consequently, the virtues of realism are best perceived against the dual background of constructivism and empiricism.[21]

An important key to the argument between realism and constructivism concerns the notion of intertheoretic continuity of reference employed above, for the typical constructivist objection to realism in any form is that once the theoretical terms of science are deprived of an independent, observational base that could serve to confirm or at least falsify their postulated reference, this reference of scientific theories become impossible to establish—particularly given the significant historical discontinuities that are said to exist between scientific theories that are only ostensibly about the same regions of reality. This argument, which should not be stated in one sentence, comes in different forms and has been given different labels. It has been dubbed, wrongly, it seems, the Quine-Duhem problem concerning a putative "underdetermination of theory by observation."[22] The absence of a continuity of reference between different

[21] Here and in what follows, see Boyd, "Current Status."

[22] See Roger Ariew, "The Duhem Thesis," *British Journal of Philosophy of Science,* 35 (1984), 313–25.

scientific theories is spoken of in terms of "radical meaning invariance," and the idea that science procedes through discrete shifts of conceptual perspective is associated with the terms "paradigm change" and "theory incommensurability." In what follows I first focus on the notions of intertheoretical incommensurability and indeterminacy of reference, as these can lead to the key issues in our context.[23] The relation between these topics and the broader issues should be obvious: if the incommensurability of scientific theories were established, one of the most basic elements of a realist demarcational strategy—namely, the possibility of science's truthful reference to mind-independent reality—would be undermined. The sciences might then be distinguished from other products of human cognition in a variety of ways, but the sciences could not be said to have discovered and explained facets of reality which were previously unknown to, or misrepresented by, other systems of belief (e.g., mythology). Only the extreme scientific realist, that is, the reductionist, wants to argue that scientific representations are the *only* ones having ultimate validity, and from such a perspective, the science–nonscience demarcation is equivalent to a knowledge–nonknowledge distinction. But moderate realism disagrees with this view, not only because moderate realism allows for the possibility of the truth of ordinary, nonscientific forms of knowledge and explanation, but also because moderate realism does not postulate the existence of a single, ideal scientific theory by means of which all properties and entities are related to an ultimate level of reality. This moderate realism *does* suggest that, when nonscientific worldviews are directly contradicted by well-justified scientific doctrines, it is not reasonable to think the former beliefs true, but this does not amount to the reductionist program. Such is the general background for the brief discussion of incommensurability to follow.

I may as well begin this discussion with the customary citation of a passage from Norwood Hanson's *Patterns of Discovery*, a work that, although published in 1958, strongly anticipated a whole trend in the philosophy of science. Taking as a central analogy the phenomenon at the border of perception and conception referred to as "aspect shift" (of which Wittgenstein's duck-rabbit is the most common example), Hanson challenges the sharpness of the distinction between visual data and their interpretation. Carried over into the domain of scientific theorizing and observation, this "observation"

[23] An intelligent discussion of these matters in relation to literary criticism is that of Trevor Ponech, "Literary Criticism and the Storm of Progress" (unpublished essay).

would imply the relativity of frameworks: "Let us consider Johannes Kepler: imagine him on a hill waching the dawn. With him is Tycho Brahe. Kepler regarded the sun as fixed: it was the earth that moved. But Tycho followed Ptolemy and Aristotle in this much at least: the earth was fixed and all other celestial bodies moved around it. *Do Kepler and Tycho see the same thing in the east at dawn?*"[24]

Hanson's answer is that Kepler and Brahe do not really see the same thing, because whatever similarity there may have been between their optical inputs was overdetermined by the essentially interpretive nature of actual seeing: "People, not their eyes, see. Cameras, and eyeballs, are blind." Thus it is theory and conceptual perspective that make the difference between what the two men see. It seems to follow that their direct visual observations cannot serve as an independent basis for testing their opposing conceptions; if observation is overdetermined by theory, theory is underdetermined by observation, and there is no theory-free evidence that can be used to confirm either of the two visions of reality. Both men can point to the "same" natural phenomenon and say "See! My theory is confirmed!"

This kind of perspective can be further illustrated with Kuhn's argument that the shift between Newtonian and Einsteinian physical theory involved a radical paradigm change, a scientific revolution realizing a "displacement of the conceptual network through which scientists view the world."[25] In fact, Kuhn suggests that this example is a privileged case of his more general conception of the nature of scientific change. Kuhn's basis for this claim is that the differences between the two theories are necessary and irreconcilable, and with this in view he sketches and criticizes an attempt to resolve these differences by means of a derivation of Newtonian dynamics from relativistic dynamics. This attempt involves adding to the relativistic laws a set of statements that restrict certain parameters so that the Newtonian laws can be interpreted as a special or limited case of the former. Kuhn suggests that it is possible in this manner, while

[24]Norwood R. Hanson, *Patterns of Discovery* (Cambridge: Cambridge University Press, 1958), 5. For a clear development of Hanson's insight within a realist perspective, see Craig Dilworth, *Scientific Progress* (Dordrecht: Reidel, 1981). Dilworth's position is easily misunderstood unless one realizes the importance of his view that beneath the discontinuities of theory there lies an essential continuity of laws. See his "On Theoretical Terms," *Erkenntnis*, 21 (1984), 205–41; and "Laws vs. Theories," in *Epistemology and Philosophy of Science*, ed. Paul Weingartner and Johannes Czermak (Vienna: Hölder-Picherl-Tempsky, 1983), 353–55.

[25]Thomas S. Kuhn, *The Structure of Scientific Revolutions*, 2d ed. (Chicago: University of Chicago Press, 1970), 101.

remaining within an Einsteinian framework, to explain why it is that Newton's laws once seemed to be accurate—the velocities in question never surpassed those specified by the limiting parameters. But this practical and approximate equivalence of observational results does not entail, Kuhn contends, a real identity of physical referents. The concepts of space, mass, and time in the two frameworks do not have the same meaning, and the Newtonian laws have been interpreted in a manner that was not possible before Einstein. Therefore the derivation of Newtonian dynamics from Einsteinian dynamics fails, for in attempting it we "have had to alter the fundamental structural elements of which the universe to which they apply is composed."[26] There is no continuity of reference across theories, and it is a mistake to think that both theories could be addressed by a single experimental result; the experiment itself would have to be designed and interpreted in function of either one of them—not both.

From here it is one short step to the conclusion that the notion of progress in science is inappropriate insofar as its application rests upon the possibility of a meaningful comparison between successive theories. But if what Newton referred to as "mass" and what Einstein referred to as "mass" are not the same things, if they do not even have the same meaning, can they really be said to represent successive approximations of a single reality? In what sense can they be compared at all? And if they cannot, must we not then abandon the notion that science as a whole converges upon knowledge of a paradigm-independent physical reality? How can we show that the local discontinuities between theories find their place upon a background of deeper continuity? If the paradigm shift between theories results in the emergence of a conception that is not about the same physical realities at all, then it is impossible to say that the one was an approximation of the other, or even that the error of the earlier theory was then corrected by its successor. Nor is it only some formalized model of theory subsumption that is threatened here, for what is put in question is our most basic and intuitive sense of the continuity of the effort and intention of science which, no longer guided by its investigation into a single reality, may as well turn upon itself in search of the social and psychological determinants responsible for its shifts in perspective. The Einstein–Newton example certainly seems decisive: it is not a question of calling in the exotic cosmologies of Babylon in order to contest the arrogant realism of Western science; rather, it seems to be a matter of the impossibility of

[26]Kuhn, *Structure*, 102.

applying a logic of justification within the very heart of modern physics, and purely on its own terms.

Is the argument compelling, then? Has this matter been settled? Are humanists warranted in citing Kuhn's book as a kind of master argument when they wish to get out from beneath the shadow of an overbearing science, unified, rational, and progressive? Robert Nozick seems to think so when he suggests that, in the light of the underdetermination of scientific theory by all possible observational data, we should think of science as an art form—one in quest of the same type of organicity prized in other endeavors.[27] I do not, however, think it is time to draw any such conclusions. It seems imprudent to assume that Kuhn's two pages on the relationship between the physics of Einstein and that of Newton could settle the matter, especially given the ambiguities of Kuhn's remarks, complexified further still by his own later commentaries. In fact, the debate goes on.

Rethinking Reference

In a few carefully argued papers, Hartry Field has challenged the inevitability of Kuhn's conclusions as well as some of the more basic assumptions about reference that may be thought to support such views, namely, W. V. O. Quine's general assault on correspondence or reference theories of truth.[28] Field does not deny that there are important instances of semantic indeterminacy in scientific theories as well as in everyday language. In the latter, for example, the expression "a tall man" is irreducibly vague, for it has no extension that can be precisely defined as a set (the boundary of its "set" is fuzzy, a matter of degree). Yet Field disagrees as to the nature of the conclusions that can be drawn on the basis of the semantic indeterminacy in science. According to Quine, the existence of indeterminacy implies that scientific terms are meaningless and denotationless except relative to their own theoretical framework, for it is only within such

[27]Robert Nozick, *Philosophical Explanations* (Cambridge, Mass.: Harvard University Press, 1981), 646.

[28]Hartry Field, "Theory Change and the Indeterminacy of Reference," *Journal of Philosophy*, 70 (1973), 462–81; and "Quine and the Correspondence Theory," *Philosophical Review*, 83 (1974), 200–28; "Realism and Relativism," *Journal of Philosophy*, 79 (1982), 553–56. Field is also the author of *Science without Numbers: A Defence of Nominalism* (Oxford: Blackwell, 1980). For additional background, see "Conventionalism and Instrumentalism in Semantics," *Noûs*, 9 (1975), 375–405; and "Tarski's Theory of Truth," *Journal of Philosophy*, 69 (1972), 347–75.

a framework that their meanings are specified; given this, it follows that there is no intertheoretical reference and that truth is immanent to the conceptual scheme. Truth, then, is an intralinguistic relation, not a relation between words and things. Quine may want to admit certain possibilities of translation and interpretation between theories, but he considers that it makes no sense at all to say that the terms of a theory denote or signify anything outside of the theory.

Field rejects this view, for he thinks that the semantic relations of denotation (of particulars or tokens) and of signification (of types) are objective relationships that hold between terms and extralinguistic objects, sets of objects, or states of affairs. Moreover, the reference relations are said not to be relative to a conceptual scheme "in any interesting sense." Hence he defends a correspondence theory of truth, but not in the caricatural form that is usually evoked in order to brush aside such positions. One key to Field's reformulation of referential semantics lies in the notions of partial denotation and partial signification, and in his claim that a sentence can have a perfectly determinate truth value while containing terms of indeterminate reference. When applied to the case of the issue of a term's meaning across scientific paradigm shifts, Field's refinement in semantics makes it possible to describe an alternate to the choices envisioned by Kuhn. According to Field, these choices are threefold: (1) the term changed denotations as a result of the paradigm shift; (2) the term, which had no real denotation before the shift, acquired one as a result of the scientific revolution; (3) the term maintained the same denotation before and after the paradigm shift. Field finds each of these characterizations to be inadequate, and prefers a fourth: (4) the term has undergone a "denotational refinement," which means that the set of things that it partially denoted after the shift in paradigms is a proper subset of the set of things it partially denoted before it, the result being the rejection of the possibility of any simple *equation* of the term's reference within the two theories, but also, a rejection of the incommensurability thesis following which their references cannot even be *compared*.[29]

How does Field's argument apply to Kuhn's example? Field takes up the problem of the meaning of "mass" in the two dynamics, inquiring whether Newton's term referred to the same physical quantity as does "mass" within the Einsteinian framework of special relativity. He agrees with Kuhn that no simple equivalence is warranted; it will not do to say that Newton's theory, with its reference

[29]Field, "Theory Change," 497.

to "mass," is a rough approximation to that of Einstein, true enough at low velocities but false in other conditions. When analyzed from the Einsteinian perspective, Newtonian "mass" reveals its *referential indeterminacy*; it can be translated into either of two claims: (1) Newton's term "mass" denotes total energy/c^2 (proper mass); and (2) Newton's term "mass" denotes nonkinetic energy/c^2 (relativistic mass). No such distinction was available within the Newtonian framework, and thus Field contends that there is no basis for telling whether Newton was referring to proper mass or relativistic mass—there is "no fact of the matter" upon which to make a decision. It is in this sense that Newton's "mass" has an indeterminate reference relative to the Einsteinian framework, for here the two possible translations are not equivalent (relativistic mass does not have the same value in all frames of reference, but proper mass does). Thus we can rule out the attempt to claim that the word "mass" kept exactly the same denotation across the paradigm shift.

At this point we encounter the issue of how this indeterminacy is to be interpreted. Kuhn chooses an option whereby the reference of Newton's terms is "Newtonian mass," whereas the reference of Einstein's term is something else entirely. Are both right, then? But that is a puzzling notion, for, if we hold Einstein's theory, we cannot believe that there exists a quantity corresponding to "Newtonian mass," and we could only hold that both were right by maintaining a wildly constructivist ontology admitting the omnipotence of thought. Is this the necessary conclusion? No, for we must also consider Field's third option, following which we would reckon that Newton's "mass" was a term having no denotation at all—or in other words, that it was a fiction having the same reference as "Santa Claus" as far as truth values are concerned. Yet there are problems with this view, not the least of which is the fact that many of Newton's utterances that contain the term "mass" are in fact true and agreed to today by relativity theorists. Field gives the following example: "To accelerate a body uniformly between any pair of different velocities, more force is required if the mass in the body is greater."[30] This proposition is true whether we take the mass to be a relativistic mass *or* a proper mass. And it is patently false if we substitute for "mass" nondenoting words such as "phlogiston" or "ghosts." This substitution test seems to reveal a nuance that was missed by the bald decision that Newton's mass was only a fiction.

Having rejected the three untenable options, Field formulates his

[30]Field, "Theory Change," 472.

own alternative. In order to capture the missing nuance, Field supplements the semantics of reference with a notion of partial denotation, and he sketches definitions of truth and falsity in terms of partial denotation and partial signification. His alternative formulation is as follows: "Newton's word 'mass' partially denoted proper mass and partially denoted relativistic mass and didn't partially denote anything else."[31] This description of the partial denotation of Newton's term allows its overlap with the Einsteinian usage without, however, equating the two; most important, the truth values are correctly conserved across the paradigm shift from the perspective of relativity, without having to say that all of Newton's propositions are simply false or without reference. Field suggests that the shift is better understood as a denotational refinement and states that such refinements are a fairly common feature of scientific revolutions. Thus we have an alternative to Kuhn's view—or rather, to the two different positions that Kuhn adopts—for if he generally holds to the complete incommensurability of the terms within the two different theories, he at one point suggests that the Einsteinian theory flatly contradicts the Newtonian one.[32]

Field's analysis is a crucial part of a general realist critique of constructivism insofar as it offers a cogent response to the contention that consecutive semantic schemes or paradigms of science are not logically commensurable. His notions of partial denotation and denotational refinement make possible a theory of reference which allows for valid judgments of univocality of reference of a theoretical term that appears in the context of different theoretical paradigms. Field's insights have been extended into a more general view of reference by Boyd, whose notion of epistemic access expresses the kernel of a realist conception of knowledge. According to Boyd, a linguistically mediated epistemic access to reality necessarily involves modification of linguistic usage by means of which language is accommodated to newly discovered features of the world. This

[31]Field, "Theory Change," 476.

[32]Field, "Theory Change," 469 n. 7. For a very different discussion of what Kuhn is saying in taking up the relation between Newton and Einstein, see Richard J. Bernstein, *Beyond Objectivism and Relativism: Science, Hermeneutics, and Praxis* (Philadelphia: University of Pennsylvania, 1983), 79–93; Bernstein argues that Kuhn, in spite of some of his rhetoric, is not really a radical framework relativist. Although he may not be, this hardly pertains to the impact of Kuhn's book. Moreover, Bernstein defends his claim by deemphasizing drastically the issue of meaning—an issue that is crucial in this context insofar as it is intimately related (but certainly not equivalent) to the question of the reference of terms across paradigms. And insofar as Kuhn's argument is supposed to have implications for the question of science's realism and for the relation of science to purely relative and historical worldviews, the continuities and discontinuities of reference remain the key issue.

process of improved access to features of reality through modification and accommodation of usage is, he states, "the very core of reference." Although I cannot in the present context rehearse the details of Boyd's theory of reference and the various arguments offered in support of it, it may nonetheless be helpful to reproduce part of his discussion before moving on: "The mark of reference, then, is epistemic acess, and the mark of epistemic acess is the relevant sort of socially coordinated epistemic success. Roughly, a general term, *T*, affords epistemic access to a kind (species, magnitude, and so on) *k* to the extent that the sorts of considerations which are (in the relevant historical context) rationally taken as evidence for statements involving *T* are, typically, indicative in an appropriate way of features of *k*." Boyd goes on in the same context to give a nonexhaustive list of a number of relations between terms and kinds which are said to be characteristic of epistemic access. These mutually supporting relations are the following:

1. Certain of the circumstances or procedures which are understood to be apt for the perception, detection, or measurement of *T* are, in fact, typically apt for the perception, detection, or measurement of *k*.
2. Some of the circumstances which are taken to be indicative of certain features or properties of manifestations of *T* are, in fact, typically indicative of those features or properties of manifestations of *k*.
3. Certain significant effects attributed to the referent of *T* by experts (or generally, in the case of nontheoretical terms) are in fact typically produced by *k*.
4. Some of the most central laws involving the term *T* are approximately true if they are understood to be about *k*.
5. There is some generally accepted, putative, definite description of the referent of *T* which is in fact true of *k* and of no other kind.
6. The sorts of considerations which rationally lead to modifications of, or additions to, existing theories involving the term *T* are, typically and over time, indicative of respects in which those theories can be modified so as to provide more nearly accurate descriptions, when the term *T* is understood as referring to *k*, so that the tendency over time is for rationally conducted inquiry to result in theories involving *T* which are increasingly accurate when understood to be about *k*.[33]

Given the coherence of this alternative possible account of reference, it seems to be pure dogmatism to assert, on the basis of Kuhn's

[33]Boyd, "Metaphor and Theory Change," 384–85. Boyd notes that he expressly abuses the use–mention distinction in the passage cited.

work, that the reference of all scientific knowledge has been shown to be wholly relative to different paradigms, or that this reference is necessarily discontinuous and incommensurable across scientific revolutions. In fact, the stronger versions of semantic relativism hardly appear to provide a solid mold into which to pour the data of the recent history of natural science. Moreover, there is good reason to assume, along with Field and Boyd, that a coherent account of scientific terms' reference across theories can be provided in many important instances. Given that this kind of alternative account remains a distinct possibility, there is no basis for going on with dogmatic assertions of the relativistic conclusions that have sometimes been drawn from the incommensurability doctrines. On the contrary, the way is open to understanding comparisons between theories in which their respective denotations can be rationally tested in terms of methods and doctrines recognized by both. More generally, it can be pointed out that the question of incommensurability, and hence of the continuity and progress of scientific explanations and theories, is itself necessarily posed in terms of certain assumptions that set real limits on the implications of incommensurability. In the case of Hanson's question about Kepler and Brahe, for example it is clear that the issue of the truth of different theoretical constructs is being raised against the background of what we have earlier referred to as the manifest image. In order to pose his question, Hanson must place the two observers "on a hill," "facing east," and "at dawn"—in other words, he situates them within a single empirical reality where they can be located together. Their two theories about the sun are framed by this context, and only if these two theories are equally tenable representations within this context can we consistently hold that the only true answer to the question is that of the aspect shift. In this particular example, this is not the case. It should also be noted that in Field's discussion, it is the Einsteinian frame that is positioned as the englobing frame of reference, without, however, this leading to the total negation of the earlier dynamic's reference to reality. This choice of frames, however, is anything but arbitrary, for reasons to be discussed below. More generally, it should be noted that talk of the radical incommensurability of theories presupposes that one is thinking within a framework where the theories are effectively compared to each other, and this comparison presupposes in turn that there are some commensurabilities.[34]

[34]This point is argued nicely by Hilary Putnam in his discussion of Kuhn in "Philosophers and Human Understanding," in *Realism and Reason: Philosophical Papers,*

In regard to the more general question of science's putative rationality and progress, it is important to make clear that the realist's theory of reference and truth is only one facet of the argument. There is a very strong argument to the effect that the real specificity of modern science's effective referential relationship to nature resides in its method of validating and invalidating beliefs and suppositions, a method the kernel of which is an insistence upon certain types of evidentiary support.[35] The notions of experiment and measurement seem to be crucial in this regard, for although both refer to kinds of observation that are clearly only possible and *relevant* in relation to scientific theory, it does not follow that the results are strongly overdetermined by any one theoretical interpretation in a broader sense. Boyd refers to this situation in terms of "pair-wise theory-neutrality of method," which involves the idea that, for two rival theories, there are experimental tests based on a method legitimized by both of them. Craig Dilworth has argued quite plausibly, for example, that singular statements about the results of instrument readings, as well as lawlike generalizations based upon them, retain a crucial degree of neutrality across theoretical interpretations of them within the developed branches of contemporary science; the same measuring operations, when carried out by people having the same competence, and under significantly identical conditions, always produce the same results within specifiable ranges of approximation. This iterability of results is not put in question by the kind of paradigm shift to which Kuhn referred, and it can, when conjoined with other realist arguments, be taken as contributing to science's reference to facets of a mind-independent nature.[36] Yet an approach that tries to define scientific knowledge in terms of its evidentiary support must find a balance between two extremes. On the one hand is the danger of expecting the scientific method to be defined as some kind of machine-like algorithm of discovery and of justification, as a

---

vol. 3 (Cambridge: Cambridge Univesity Press, 1983), 184–204. See also his fine remarks on cultural relativism in "Why Reason Can't Be Naturalized," 229–47. Putnam, however, critiques what he calls "metaphysical realism" and does not think that the self-destroying nature of relativism lends any support to this kind of realism, which requires a correspondence theory of reference and truth. For Putnam, truth is an idealization of rational acceptability, and reference is fixed only relative to a semantic framework. Many different correspondences can satisfy the constraints set by a given language, and no "God's eye" view can be had by means of which to check these correspondence relations against an external reality. For criticisms, see Field's "Realism and Relativism"; and Millikan, "Metaphysical Anti-Realism?"

[35]Harvey Siegel, "What Is the Question Concerning the Rationality of Science?" *Philosophy of Science*, 52 (1985), 517–37.

[36]Dilworth, "On Theoretical Terms," 410–11.

perfect epistemic automaton (rationality equals infallible calculation). When it turns out than none has ever existed and cannot be designed, the conclusion often drawn is that there is no rationality in the scientific method. On the other hand is the danger that the appeals to the superior rationality of scientific method remain wholly vacuous because nothing more specific is said about what this method is, or about how it yields reliable evidence. Can such a balance be achieved?

On this vast issue I must limit myself to a few remarks, my goal being, once more, the modest one of simply evoking the existence of arguments and positions that are too frequently ignored in humanistic and literary contexts. I take as my guide the assertion of Isabelle Stengers and Ilya Prigogine that "the experimental procedure defines the set of dialogues with nature attempted by modern science; it is the basis of that science's originality, specificity, and limits."[37] These authors point out, in a Baconian vein, that the kind of experimental dialogue that characterizes modern science has nothing to do with a passive or neutral observational stance; nor is it marked by a general respect for the wealth of empirical details. On the contrary, the experimental approach is a practical intervention whereby an ideal situation is constructed, a situation that brings a natural reality into maximal approximation to theoretical description precisely in order to test the limits of such an equation. Thus the method of experimental dialogue amounts to a highly selective and manipulative *mise en scène* of nature. It is like a courtroom scene, in fact, one in which the judge poses very specific questions to the witness, questions having sense only in function of more general principles and a theoretical background. The judge listens carefully to the responses, which *are* those of nature, for even if these responses are only those of a prepared and simplified nature, they are still capable of giving clues about the reliability of the hypotheses and suppositions behind the questions. Even in such restricted conditions, nature retains its power to say "no"—although it is most often up to the scientist to conjecture about the precise reason why the expected result did not occur.[38] It is important to point out, with Boyd, that

[37]Ilya Prigogine and Isabelle Stengers, *La Nouvelle alliance: Métamorphose de la science* (Paris: Gallimard, 1979), 48. See also Mario Bunge, *Treatise on Basic Philosophy*, vol. 6: *Epistemology and Methodology II: Understanding the World* (Dordrecht: Reidel, 1983), 59–113.

[38]According to Roger Ariew, this is the real thrust of Pierre Duhem's "separability" and "falsifiability" theses, which concern not "science in general" but the situation of an advanced and highly theoretical physics where certain hypotheses do not have

the experimental method of science relies heavily upon inductive inferences. These inferences have a lot to do with the scientist's judgments about which observational findings are to be expected and which would be surprising. The ceteris paribus clause, which is necessary to the isolation of an autonomous experimental context of analysis, can only be satisfied by virtue of this kind of inductive inference, which is made on the basis of relevant background theories. As Boyd notes, "in the typical case in which a scientist tests a theory by testing one of its observational predictions, the prediction in question will have been inductively rather than deductively derived from the theory in question together with relevant auxiliary hypotheses."[39]

The experimental dialogue, then, is "theory bound," as the constructivists would indeed have it, but it is not clear that the constructivist epistemology is correct as a result. Moreover, it is important to agree with the sociological reductionists that the experimental method is also a conversation between human agents, a conversation conditioned by specific kinds of institutions. But again, it is not clear that the relativist's conclusions follow from this truth. On the contrary, it is possible that the *specific* institution and tradition-bound character of scientific methodology is what makes possible and enhances referential reliability. Given certain dialogical norms, only on a short-term basis can the individual scientist (or group of them) project onto nature only those responses they want to hear. For there are others, competitors and collaborators, who will pose the same questions and who will be quick to contradict any projections. More generally, then, it is important to note that certain types of social systems are necessary conditions to scientific inquiry.[40] In the context of contemporary literary theory, it may also be useful to point out in passing that the institution of writing is a privileged instrument of scientific work, for it is by means of transcriptions that data and conjectures can best be retained for the public scrutiny essential to the elimination of idiosyncratic inter-

---

direct, easily isolatable empirical consequences; as a result, even when an experiment seems to disconfirm a hypothesis, it will not give a specific indication as to which aspect of the theory is erroneous; "Duhem Thesis," 318–23.

[39]Boyd, "Logician's Dilemma," 243–44.

[40]For a discussion of some of the social systems implications of Popper's methodological rules, for example, see Paul Dumouchel, "Social Systems and the Logic of Discovery (1)," *Cahiers du C.R.E.A.*, 5 (1985), 95–125. As these *Cahiers* are not widely distributed and will be cited several times in what follows, I add the address: CREA, 1 rue Descartes, Paris 75005.

ventions in the dialogue.[41] Briefly, then, the experimental dialogue may be characterized as an institutionally organized set of practices resulting in a process of inquiry that is critical, progressive, and self-correcting.

Has my discussion at this point successfully established the minimal claim announced at the beginning of the chapter? I have held open the possibility that scientific observations—be they as theory-laden and socially conditioned as one likes—sometimes achieve a "partial denotation" that remains consistent across theories. This consistency of reference provides the basis for an experimental/observational/inferential comparison of the relative explanatory strengths and weaknesses of competitive theories; in other words, it provides for the possibility of evidential bases for decisions as to the relative justification and falsification of alternative hypotheses. Such a preliminary defense does not amount to a detailed discussion of the nature of scientific explanations; it does not take up, for example, the thorny issue of the status of causality and the debate between regularity and realist accounts of lawlike explanation (more on patterns of scientific explanation in Chapter 6). Nothing has been said about *which* theoretical terms of science really refer to natural entities and which do not. It should be noted that Rom Harré, one of the staunchest defenders of a realist conception of science, holds to the necessity of such a distinction. But if in some cases the entities do not exist, this reservation is no help to the constructivists, who must do away with all reference to nature, and who are committed to the view that all of science's theoretical terms are constructions of reality.[42] But this constructivist position flies in the face of the extensive historical evidence that unobservable and latent entities postulated in scientific theories have been central building blocks in successful explanations and predictions.[43] Current scientific prac-

[41]Gilles-Gaston Granger, *Formal Thought and the Sciences of Man* (Dordrecht: Reidel, 1983), 32–39.

[42]Rom Harré, *Theories and Things* (London: Sheed & Ward, 1961); for historical background, see Mary Jo Nye, *Molecular Reality* (New York: Elsevier, 1972); cited and discussed by Wesley C. Salmon, "Comets, Pollen and Dreams: Some Reflections on Scientific Explanation," and "Further Reflections," in Robert McLaughlin, ed., *What? Where? When? Why?: Essays on Induction, Space and Time, Explanation* (Dordrecht: Reidel, 1982), 155–78; 231–80.

[43]For valuable remarks on the relation between historical evidence and the arguments surrounding a realist interpretation of theoretical terms, see Boyd, "Current Status," 87–89. A skeptical view is that expressed by Isabelle Stengers in "L'Histoire des sciences et comment s'en servir," in which she proclaims that "the history of science is a struggle between thieves in which each interprets the past in his own way, and where the past is itself a series of discourses where it was a matter of interpreting the past"; in Jean-Pierre Dupuy, Félix Guattari, Bruno Latour, et al., *Sens et place des connaissances dans la société* (Paris: C.N.R.S., 1986), 115–50, citation: 149.

tice as well as important parts of the historical record support the empiricist view that it is incorrect to draw a sharp dichotomy between observation and theory whereby it is only reasonable to believe in the former.[44]

As an example, we may evoke the case of Harvey's theory of the circulation of the blood, a discovery that boldly contested the institutionalized conception.[45] As is well known, Harvey's theory led him to postulate the existence of a link between the arteries and the veins, a link that was for him an unobservable, theoretical supposition, but that became an intersubjectively tested and observed fact after Malpighi, using an early microscope, saw and identified the capillaries. If Harvey's theory was a construct, he should be credited with a puissant and far-reaching constructive power, for people are now able to use microscopes to observe the corpuscules of blood as they move through the capillaries from the tips of the arteries to the veins.[46] And although the present medical science is greatly removed from some perfect or total knowledge, to imagine that Harvey's discovery will someday be totally negated by a different theoretical paradigm is to engage either in an incoherent or in an irrelevant form of skepticism.[47] What *is* coherent is to suppose that Harvey's work

[44]A historical example that is *not* taken up here is that of quantum mechanics, which is for many literary thinkers the decisive case upon which the evil "positivist" philosophies of science are to be refuted. I merely point out to those inclined to be persuaded or intimidated by such references that the Copenhagen interpretation of quantum mechanics is one interpretation among others, and that the realistic approaches to it have not yet been exhausted or refuted. It is possible, for example, to put the problem concerning the wave–particle interpretation into the compartment of quantum measurement theory and then claim that quantum theory itself says nothing about either particles or waves. A clear and nontechnical discussion can be found in Mario Bunge, *Causality and Modern Science* (New York: Dover, 1979), 346–53. A detailed and technical realist approach is given in Raimo Tuomela, *Science, Action, and Reality*, 52–64.

[45]This example was first brought to my attention by René Girard.

[46]W. M. O'Neil, *Fact and Theory: An Aspect of the Philosophy of Science* (Sydney: Sydney University Press, 1969), 5–14; the detailed case studies presented in this work are a useful antidote to some of the more speculative and irrationalist types of philosophy of science. Other examples are presented in Wesley C. Salmon, *Scientific Explanation and the Causal Structure of the World* (Princeton: Princeton University Press, 1984), 267–79. Attempts to stress the less modern and progressive aspects of Harvey's thought have been usefully criticized by Brian Vickers in the introduction to his anthology, *Occult and Scientific Mentalities in the Renaissance* (Cambridge: Cambridge University Press, 1984), 13–15.

[47]An excellent article that may lend me some support is Alison Wylie, "Arguments for Scientific Realism: The Ascending Spiral," *American Philosophical Quarterly*, 23 (1986), 287–97; the author argues convincingly that the dispute between realism and antirealism quickly elevates to being a matter of basic epistemological commitments, the nonrealist typically being committed to diminishing all epistemic risks, the realist arguing that we are already committed to such basic principles as abductive inference and cannot question them without taking an incoherent or irrelevant stance.

achieved only a partial reference to the realities of the blood within the human body, and that as such, it is open to future refinement. It should go without saying as well that no one can make the kind of observations in question here without relying upon the proper theoretical and practical background knowledge; what does not follow from that, however, is that it is the background of theory which somehow magically produced an illusory circulation of the blood which, prior to Harvey, did not exist. Were human bodies populated by the four humours and so on until the seventeenth century?

Here we return to Boyd's key argument against the antirealists, the argument that allows the typical anticonstructivist retorts I have just been rehearsing to fall into place within a general, pro-realist conception. As Boyd points out, although an epistemic-access account of reference and a defense of pairwise theory neutrality of method are both important parts of a realist conception, they do not suffice to refute the key epistemological claim made by the constructivist. In Boyd's view, this epistemological claim is the belief that scientific methodology is so totally theory dependent that it must be seen as a procedure of constructing, not discovering, realities: "What the constructivist argues is that a general methodology which is predicated upon a particular theoretical tradition, and which is theory-determined to its core, cannot be understood as a methodology for discovering features of a world which is not in some significant way defined by that tradition."[48] The world is largely defined or constituted, then, by the theoretical tradition informing scientific methodology.

What is the realist answer to *that* most basic contention? It cannot return to any of the props that supported the empiricist's various unsuccessful attempts at solving the demarcation problem—such as the myth of theory-neutral observational languages, protocol sentences, or a wholly deductive logic of confirmation and falsification. Boyd's answer is to point out a crippling weakness in the constructivist's vision, namely, the fact that constructivists cannot explain the instrumental reliability of scientific theories. And to this weakness in the constructivist account corresponds the positive hypothesis defended by the realist conception, namely, a highly plausible explanation of the indubitable instrumental reliability of scientific theory: "The reliability of theory-dependent judgments of projectibility and degrees of confirmation can only be satisfactorily explained on the assumption that the theoretical claims embodied in the back-

[48] Boyd, "Current Status," 59.

ground theories which determine those judgments are relevantly approximately true, and that scientific methodology acts dialectically so as to produce in the long run an increasingly accurate theoretical picture of the world."[49] The point here is that the realist has presented an argument to which the constructivist cannot easily respond. The constructivist must either offer an alternative explanation of science's instrumental reliability or deny the very reality of this reliability. Yet the latter option entails a serious inconsistency if the constructivist goes on flying in airplanes and no plausible constructivist account of instrumental reliability exists. The possibility that constructivists cannot allow is that some of the languages of science manage to realize a partial and abstractive, but accurate and true, *reference* to aspects of natural reality. Instead, constructivists would have us believe another, unspecified explanation of how a fictive, wholly discontinuous, and irrational science goes on achieving its many successful interventions in nature and in society, including the most catastrophic possibility of all, the destruction of the human race. *Ignotum per ignotius.*

On the Difference between Science and Scientism

It is difficult to discuss science with students of literature without quickly getting the impression that one is engaged in a *dialogue de sourds.* It seems to me that a prevalent feature of the presuppositions brought to such discussions is the organization of arguments and evidence in terms of a single false choice. Put bluntly, either one participates in the counter-Enlightenment, romantic, symbolist, and liberationist denunciation of science as a whole, or one is engaged in the loathesome business of offering justifications for the very thing that the former camp detests. But if I want to insist upon anything, it is that this false choice is based upon a myth, and there are other, more plausible and worthwhile positions to be seriously entertained so that we can move beyond such sterile alternatives.

What, then, is this myth? Put simply, it is the idea that science equals scientism. Scientism in this context is an incoherent constellation of erroneous and dangerous views, beginning, for example, with the notion that science is a value-free and neutral type of inquiry. Scientism furthermore holds that science is some kind of supreme good, valuable always and in itself, and that technology and

[49]Boyd, "Current Status," 65.

applied science are contingent appendages to this purely speculative curiosity. In its most extreme forms, scientism becomes a religion, one that pretends to offer us "the whole picture," the "meaning of it all." Another scientistic mistake is to think that the scientific attitude dictates our belief in a naturalistic and physicalist reduction of the psychic, social, and historical levels of description to some set of more basic, materialistic terms. Thus it is scientism that drives a wedge between the so-called two cultures and casts the humanities into the shadows of a glaring, omniscient Scientific Knowledge.

The ultimate crystallization of the scientistic philosophy can be found in the metaphor of Laplace's demon. This otherworldly entity observes the state of the universe from without, and given a knowledge of its laws coupled with a perfect description of one of its instantaneous states is capable of calculating in perfect detail all of the universe's states—past, present, and future—as if time made no difference whatsoever. Need we add that the attributes of the states in question are entirely reducible to those of matter located in space, as in Alfred North Whitehead's complaint about the universe of the mechanistic philosophy: "A dull affair, soundless, scentless, colorless, merely the hurrying of matter, endlessly, meaninglessly."[50]

My point in listing these tenets of scientism is to state that none of them is entailed by our critique of framework relativism and constructivism; none is entailed by the moderate form of realism; none is either supported or required by an understanding of the realistic position under discussion. In fact, I have contended that science is a selective and highly oriented attitude toward nature, and that it is neither always the best nor the only such attitude. It *is* moreover an attitude motivated by certain values, the foremost of which, is *truth*. Karl-Otto Apel believes it possible to derive from this value commit-

[50]Alfred N. Whitehead, *Science and the Modern World* (New York: Free Press, 1967), 54. An excellent discussion of the implications of the metaphor of Laplace's demon, and of the reasons for its inapplicability to real science, is to be found in Karl R. Popper, *The Open Universe: An Argument for Indeterminism*, ed. W. W. Bartley III (Totowa, N.J.: Rowman and Littlefield, 1982). My one objection to Popper in this regard is that he speaks as if Laplace thought such an achievement were really possible: "Laplace's demon is not an omniscient God, merely a super-scientist. He is not supposed to do anything which human scientists could not do, or at least approximately do: he merely is supposed to be able to carry out his task with superhuman perfection" (30). Yet compare Laplace's own discussion, which includes the remark that human scientists will "forever remain infinitely remote from such an intelligence"; Pierre Simon de Laplace, *Théorie analytique des probabilités*, 2d ed. (Paris: Courcier, 1814), iii. For Laplace the model of knowledge represented by the demon is a kind of vanishing point or ideal which the scientist should strive to approximate, but which is forever out of reach.

ment of science other "oughts" that form an indispensable ethical framework or a priori of succcessful communication and inquiry—I do not pursue this here.

Is it enough to say that science's guiding value and orientation is specified by the goal of truth? Not at all, for this says nothing about which kinds of truths science tries to produce. Nor can we quickly solve the problem by saying "explanations," for not only do we require a theory of explanations, but science surely does not set out to explain every little thing. The orientation of science toward the real is much more specific than that. This orientation—or to use the Habermasian terminology, "interest"—that is wired into scientific inquiry is often referred to as that of "instrumentality."[51] If science's orientation toward truth is very fundamental, the more specific directions in which its inquiry tends have to do with an additional attitude; the orientation toward truth alone does not explain, for example, why science focuses not on single events, but on types of events; not on particularities, but on classifiable, countable, and measurable phenomena. Nor does the goal of truth explain why science is most interested in lawlike regularities, in knowing under what conditions some kinds of things tend to happen and what other kinds of conditions and events could make them happen in a reliable and regular way or prevent them from happening with some assurance. Finally, the open-ended goal of truth cannot explain why much of scientific research is focused on the resolution of the anomalies discovered within the existing body of scientific knowledge.[52]

"Instrumentality" may be an appropriate global term for this cluster of attitudes that have characterized science's specific mode of inquiry, its orientation toward a certain type of truth. Another expression of this attitude is modern science's overriding interest in an *efficient causation* within a resistent and lawful natural reality. Some thinkers hold that science's interest in causation begins with mankind's sense of his own active, causal potentials, that it starts with the observation that one can intervene and make changes in

[51]Jürgen Habermas, *Knowledge and Human Interests* (Boston: Beacon, 1971). See also Thomas McCarthy, *The Critical Theory of Jürgen Habermas* (Cambridge, Mass.: M.I.T. Press, 1978). For a presentation of Habermas's more recent views in English, see his "Remarks on the Concept of Communicative Action," in *Social Action*, ed. Gottfried Seebass and Raimo Tuomela (Dordrecht: Reidel, 1985), 157–78; see also in the same volume Ernst Tugendhat's incisive, and in my view telling, response, "Habermas on Communicative Action," 179–86.

[52]A useful discussion of this latter point, along with a clear presentation of its relation to the deductive-nomological model of explanation, is given in Willard C. Humphreys, *Anomalies and Scientific Theories* (San Francisco: Freeman, Cooper, 1968).

reality—but not always, and not in just any way. Yet it can also be plausibly argued that it is not only human beings that act as causes in the physical world, and that efficient causation is nature's way as well, on clear display for all to see. But if "instrumentality" is a good label for the underlying orientation of scientific inquiry and intervention, one must still resist the temptation to accept the negative connotations that instantly come to mind, even if these connotations do in fact come to mind for a lot of far-reaching empirical reasons, not the least of which is science's enslavement to the thanatocracy. That the instrumental attitude as such is not inherently pernicious to the human good or to emancipation can be suggested (but of course not proven) by means of a single observation. It strikes me that those who castigate science's presumed instrumental and dominating attitude toward nature are necessarily using the concept of nature in a normative rather than descriptive manner; nature is "the good," it or She is "in order" as things are, and it is mankind's acquisitive and instrumental attitudes that initiate violence. But does not this assumption merely perpetuate the myth of the Garden? A less mythical, yet still extremely prescientific description of nature is clearly much less bucolic insofar as any human interests, instrumental or other, are concerned. This is not an argument for the neutrality of any specific modern technology; it is a rejection of the idea that in the relationship between human beings and their natural environment, an instrumental stance is necessarily aggressive and domineering (or morally bad). Indeed, the instrumental stance can also be defensive and protective, and it assures that other goods find their conditions of possibility. In the absence of an accurate and effective knowledge of some of the natural order's regularities and particularities, in the absence of human civilization's shelter against the violences of nature, it is human life that would be a "dull affair," painful, brief, comfortless, and lacking in both grace and thought. The question, then, of the interest of science's truths is usefully raised by the notion of an instrumental attitude, but this concept does not offer a reliable global viewpoint on the difficult problem of the moral and political values of actual examples of scientific knowledge.

PART THREE

THEORY AND THE QUESTION OF THE HUMAN SCIENCES

In the next two chapters I take up some of the major arguments in favor of a position that is sometimes loosely referred to as "methodological dualism." Most generally, what is at stake here is the idea that the basic kind of inquiry and explanation characteristic of the modern natural sciences is not applicable to the realm of "the human"—or, in more moderate versions, to the important aspects of human experience. According to this kind of viewpoint, insofar as there can be an adequate and appropriate knowledge of these human realities, it is achieved by methods or approaches differing fundamentally from those of the natural sciences. The pertinence of such a viewpoint in the context of the theoretical debates over literary knowledge, and over the role of theory therein, should be obvious. A critic may very well grant that the natural sciences produce a certain kind of explanation and have thereby achieved a certain type of progress. Yet the very same critic may add quite consistently that no such goals may be achieved by any of the humanistic disciplines. Thus it is crucial to address some of the major arguments supporting such a stance before moving on to the question of literary knowledge per se.

Methodological dualism is a matter of a variety of related conceptions—indeed, such positions are so numerous and complex as to be bewildering. Various traditions conspire to inform us of the different ways the objects and methods of the human and natural sciences diverge.[1] An important vein, for example, is that associated with

[1]A useful survey and criticism of arguments against objectivity in the social sciences is provided in Frank Cunningham, *Objectivity in Social Science* (Toronto:

neo-Kantianism in nineteenth century German philosophy.[2] For the neo-Kantians, one of the central differences has to do with the role of values and singularity within the human sphere; whereas the natural scientist focuses on broad regularities, recurrent types or patterns of events and their lawful repetition, the historian and critic must seek to capture the uniqueness of the phenomena they study. The humanists strive to describe rather than to explain, to list and delineate rather than to subsume and generalize. Another typical remark adds that the human, because of its creativity and freedom, is precisely *not* a determinate nature governed by lawful necessity and thereby escapes from science's net. The concepts of the humanities must, as a result, be open ended, unlike the "closed" definitions and nomic generalizations applicable to a material nature that "has no history." From here one may go on to add that, while it is the nature of material realities to be susceptible of being measured and counted, only qualitative determinations can be applied successfully to the sphere of the essentially human. Furthermore, it can be argued that language, which is often taken to be the very emblem of the anthropological difference, is irreducible to the orderly logic of sets and identifiable items; it is ultimately a matter of an irreducible and vast dimension of figures and associations, of tropic referrals, displacements, and condensations defying the rules of classical logic and with it the entire *ratio* of scientific inquiry.[3] While human beings are held to escape from the purview of science by virtue of language, they also are thought to do so by virtue of the special nature of their volition. Whereas nature is a matter of more or less lawful *events*, people engage in *action*, which cannot be calculated or plotted in the same way as events are explained. Rather, action must be understood in an interpretive movement that is essentially part of a process of

---

University of Toronto Press, 1973). Although objectivity, as Cunningham defines and defends it, is clearly an essential feature of authentic inquiry, an examination of this conception does not exhaust the question of the natural science–human science dichotomy.

[2]Apparently it is possible to identify seven different strains of neo-Kantian thought in nineteenth-century Germany, but such details are not essential to the present context. See Herbert Schnädelbach, *Philosophy in Germany, 1831–1933*, trans. Eric Matthews (Cambridge: Cambridge University Press, 1984), 106. The kind of position I have in mind is exemplified fairly well by Heinrich Rickert, *Kulturwissenschaft und Naturwissenschaft* (Freiberg: J. C. B. Mohr, 1899).

[3]This is a position presented by Cornelius Castoriadis in *L'Institution imaginaire de la société* (Paris: Seuil, 1975), esp. 324–54. For a concise presentation of similar points, see his "The Imaginary: Creation in the Social-Historical Domain," in *Disorder and Order: Proceedings of the Stanford International Symposium, Stanford Literature Studies*, 1, ed. P. Livingston (Saratoga, Calif.: Anma Libri, 1984), 146–61.

communication. Moreover, given that the natural scientist does not really communicate with her objects, her knowledge of them cannot give rise to the kinds of "Merton effects" that a social and human knowledge can produce; the social scientist who publishes his theory of social reality may, in the same stroke, be the catalyst of a change in that reality, for the objects of his study may behave differently as a result of learning about the scientist's theory. Sociological knowledge, then, can be a self-fulfilling prophecy, and no such phenomenon is known within the natural sphere.[4] Another dualist notion contends that the difference between the humanities and the sciences has to do with the very basic attitudes and orientations that underlie them. These orientations—or "knowledge-constitutive interests," to employ Habermas's expression—have a lot to do with the selection of a field of research, with its organization, and with the kind of knowledge of it that is sought. As we concluded our last chapter with a discussion of the instrumental interest that seems to be essential to the orientation of the natural sciences' perspective on nature, it is here necessary to place a great deal of emphasis on similar considerations pertaining to the basic orientations of the humanistic disciplines. Such thinkers as Habermas and Apel believe that it is crucial to be cognizant of essential differences between the interests of the natural and human sciences, for the correct orientation of the latter is held to be essentially communicational, critical, and emancipatory.

This brief overview of these dualist *topoi* should not leave the reader with the impression that such positions are always taken by those on the side of the humanities, or that such boundaries are always erected in order to defend humanity from the incursion of a violent and dominating scientific method; rather, these dualisms are also the work of philosophers who want to tidy up the sphere of science by expelling from it those troublesome realities that offer too much resistence to their procedures. It may well be that one of the tenets of positivism is an imperialistic desire to extend the natural scientific method into the domains of psychology and sociology, but it should be noted that positivists tend to understand the latter only in a very special and selective sense. Much of what falls within the

[4]This point is stressed by Karl-Otto Apel, "Types of Social Science in the Light of Human Interests of Knowledge," *Social Research*, 44 (1977), 425–70; and by Anthony Giddens, *Studies in Social and Political Theory* (New York: Basic Books, 1977), 12. A plausible rejoinder is that of Adolf Grünbaum, "Historical Determinism, Social Activism and Prediction in the Social Sciences," *British Journal for the Philosophy of Science*, 7 (1956), 236–40.

humanistic sphere is defined negatively and is posited as a residual domain of chaotic and irregular "values," "subjective impressions," "figurative expressions," "phenomenal illusions," and "secondary qualities." It is the sphere of the inessential, the contingent, the nondeterminant, the irregular and the haphazard, none of which can readily be made the object of a properly predictive, explanatory, and systematic knowledge.

Depending upon the way these various methodological and ontological differences are construed, the relation between the two main families of knowledge can be perceived as a conflictual opposition, as a total incommensurability, as complementarity, or as a form of integrated and systematic differentiation. And the discussion of the relations between the so-called two cultures, or between the "hard" and the "soft" sciences, will remain hopelessly confused, indeed pointless, until the deep assumptions underlying these differences are singled out. Of course the supposed unity of something called "the human sciences" should not be taken for granted in advance; an analysis of methodological and ontic differentiations *within* the anthropological sphere could serve to displace the terms of the dualisms in a variety of ways. One could, for example, hold to the unity of the natural and social sciences while being a dualist on the question of their relation to something called "humanistic" or "cultural" knowledge. Briefly, although one is most likely to concede that human beings and societies can be counted and measured in regard to some features that are indeed pertinent to their existence (which is what actuaries in fact do), it may at the same time be held that this body of calculations, although it might qualify as social science in some sense, does not address what is most significant, meaningful, or essentially human in individual and cultural life. Another approach entirely is needed if we are to fathom those depths—or so the argument runs. It is in regard to this kind of position that some of the most significant debates over the status of literary knowledge occur, for literary critics typically assume that their interest should lie in what is most singular, spontaneous, and atypical about the expression of conscious human experience, and not its supposed governing regularities, causal conditions, and recurrent patterns.

In the light of this variety of positions, it is clear that the various dualist stances cannot be adequately labeled with the notion of "methodological dualism." The kinds of dichotomies in question are not purely methodological, insofar as the fundamental differences, far from being a matter of specific methodological procedures, ultimately arise from more basic ontological tenets and epistemological principles. How, then, can we begin to come to terms with this diversity of viewpoints? Given the extreme variety and complexity of the issues at hand, it might seem that the only way to proceed is to get down to the specifics, attempting to untangle the various knotty subproblems related to the general question of the unity of the sciences. One could begin, for example, by taking up the problem of the applicability of the "experimental method" in such disciplines as sociology and history. Suppose, then, that one could discover reasons for which the conditions of valid experimentation (its philosophical and statistical principles) could not be satisfied in relation to crucial aspects of the anthropological domain; moreover, let us suppose that those methods that have been proposed as analogues in the human sciences (such as the comparative method suggested by Durkheim in the final chapter of his *Rules of Sociological Method*) be judged finally as not capable of serving a function analogous to that of experimentation.[5] It seems to follow that, insofar as experimentation is held to be one of the key features of the natural scientific method, one is led to decide, at the very least, upon some kind of methodological dualism based upon the difference between the specific procedures of inquiry (and underlying principles) within the different disciplines. It could be decided, for example, that while the natural scientist can conduct experimental observations using instruments and techniques of measurement, the humanistic scholar only obtains evidence of his or her object domain by means of a fundamentally different process of reading or interpretation (and here a hermeneutical version of framework relativism would come into the foreground). This dualism could then have various consequences for our sense of the validity and nature of our knowledge of human beings; it could lead to the espousal of some specific ontological theses about the ways the mode of being of natural entities differs from that of such objects as psyche, history, and society.[6]

In spite of the importance and interest of the issue of experimentation, and of others like it, I do not think that the argument can correctly begin in the manner just sketched. It seems to me that one very major and basic question has been begged in a discussion that

[5]Emile Durkheim, *Les Règles de la méthode sociologique* (Paris: Alcan, 1895), chap. 6.

[6]Readers who are inclined to conclude along these lines may profit from the alternative considerations set forth in H. S. Bertilson, "Methodology in the Study of Aggression," in *Aggression: Theoretical and Empirical Reviews*, 2 vols., ed. Russell G. Geen and Edward I. Donnerstein (New York: Academic, 1983), I:213–45.

begins with an examination of the applicability of experimentation and then moves on to a discussion of the anthropological difference as such. This question concerns the methodological and ontological assumptions that underwrite the entire inquiry and produce an argument in favor of a natural science–human science dualism. The question is simple: what kind of knowledge warrants this dualism? If the dualist's characterization of the types of knowledge is held to be exhaustive, then it is self-applying, in which case the question can be further specified: is the knowledge that warrants a natural science–human science dualism itself a natural or a human science? This rather interesting question is all too often dodged or ignored by methodological dualists, who trace limits and compare methodological differences without ever once discussing the implications of their dualism for their own demarcational labors. Yet the question that is in this manner ignored is crucial to such arguments, for the very status and validity of the methodological demarcations hinge on the legitimacy, justification, and status of the knowledge that draws the boundaries between methods and disciplines. In fact, although this metalevel question may not be explicitly addressed, there is always an implicit answer to it in the epistemological *and* ontological assumptions that are made by *methodological* dualists. Implicitly they must answer whether their conclusion in favor of a dualism of methods between the natural and human sciences is dictated by the former or by the latter. Furthermore, in order to be thoroughgoing dualists, they must hold that it is the human science that specifies and justifies its own difference. Otherwise the methodological differences held to be essential would turn out to be regional differences discovered by and noted within an englobing natural scientific conception. The radical dualists, then, can be coherent only if they hold that their entire dualist discourse relies upon an approach that differs significantly, if not fundamentally, from the methods and epistemological principles of natural science; it is not that scientific inquiry, in its encounter with the specificities of the human domain, traces the limits of the applicability of some of its techniques and procedures and moves on in search of others that may achieve similar ends; rather, the inquiry must begin with some type of cognition or set of knowledge criteria that are in some way fundamentally different from those of natural science. These would be the terms by which one judges the inapplicability of natural scientific models and tools to such realities as religion and art, for example. For this reason it is not correct to begin the discussion of the radical dualisms with the specific problems pertaining to experimentation or any other particular aspect of natural-scientific inquiry and explanation; nor will it do to start with the problems of language, metaphor, or meaning, with value and interpretation, or with any of the other material and epistemic features which in an apparently immediate manner propose themselves as candidates for the decisive anthropological difference. Prior to that, we should find out which court the ball is in, and what kinds of assumptions have already established the rules of the game. In regard to this issue, the single most crucial point concerns the image of natural science that figures in any such set of assumptions. This latter consideration is crucial because the different starting points for the discussion are *not* all symmetrical, and getting clear about the context of the discussion can help us to distinguish between major and minor disagreements, between real and false debates.

There follow from these points certain insights concerning the search for a viable strategy of dealing with the immensely difficult question of the unity of the sciences, a question that surely cannot be settled, here or anywhere else, solely by means of general theoretical and conceptual argumentations. But if "pure theory" alone cannot settle the matter, it may very well be an adequate tool for dealing with what theory itself has wrought. Although I do not think it possible or even desirable to try to lay down a set of detailed theses about the correct nature of the human sciences—among which would figure a science of literature—I do think it possible and worthwhile to present some general and theoretical arguments against those theories that rule out, in advance and as a matter of principle, the possibility of arriving at an adequate and realistic science of man. And this argument, as I have just suggested, should begin by asking what is presupposed in such dualist principles. I also think that these very general theoretical issues have direct consequences for even the most particular and detailed research, but I seek only to illustrate those consequences in subsequent chapters, where some of the specificities of literary and humanistic research are taken up directly. Yet the reader, who has once more heard the words "bear with me," should have an objection: once again, the author has shifted the burden of proof off his own shoulders and onto those of the other camp. If the dualists cannot provide an invulnerable account of their position, it is said, we are expected to imagine that the opposite viewpoint is right, even though it has neither been specified in detail nor positively defended. Is this not unfair and one-sided? The response is that the situation is indeed one-sided in a sense, but that it is not unfair as a result. There seems no good reason to believe that the two basic positions or attitudes in opposition here are in fact purely symmetrical. On the one hand, we have a viewpoint that holds essentially that, for reasons of principle, there can be no science of man and culture. (Whether there *should* be one, whether it is desirable, is another issue, which should be taken up separately.) On the other hand, we see the following reasoning: the efforts of natural scientists have, in a relatively short time (given the scope of human history), resulted in a very impressive (albeit incomplete) knowledge of aspects of natural reality; furthermore, science is not *wholly* mute on the subject of human and social realities; given this, why should we conclude in advance that the extension of this same effort can yield no further results in regard to the phenomena associated with the human psyche, with history, society, and culture? These, then, are the two basic stances, and my point is that they are not really on an even footing today. Thus the burden of proof effectively belongs on the side of those who, as a matter of theoretical principle, wish to trace a limit to a successful enterprise (and I must insist once more that to use the word "successful" here does not imply the political or moral goodness of the sciences). Moreover, this objective asymmetry is one reason people opposed to the idea of a human science frequently prefer to shift the argument to a critique of realist conceptions of the latter. Should it in fact be possible to show that the natural sciences are purely relative, irrational, and lacking in progress, then it would indeed follow that there is no asymmetry, and one would not have to fear a positivist invasion within the humanities. Yet, as I have already argued, no such demonstration has been made. Rather, the achievements of the natural sciences, when they are understood in terms other than those of the positivist and Laplacean mythologies, stand as a viable model for work within the anthropological disciplines. For this reason, we are warranted in asking the opponents of this alternative vision of the unity of the sciences to justify their claims for a radical barrier or limit to its applicability, and we want to know what must be assumed for such a barrier to be defended consistently.

My point is certainly *not* that all specific methodological and ontological problems can be settled once and for all by shifting attention to the question of the theoretical metalanguage and its implicit and explicit assumptions—indeed, it would be hard to imagine a more dogmatic and unreasonable notion. The point is rather that this shift of attention can reveal the emptiness and incoherence of some of the prominent stances on the ways humanistic knowledge must, *as a matter of principle*, differ from that of the natural sciences.

(Whether the social sciences have in fact "found their Newton," as the saying goes, is certainly not the issue to be settled here.) Bluntly put, my contention can be stated as follows: talk of the abyss that separates the "two cultures" is totally misguided insofar as one of these cultures is in fact already situated within a space defined by the other culture. In other words, the institutional and conceptual space of "knowledge" is today already organized in function of concepts and motives of the model of natural science. Thus the natural-scientific attitude effectively stands as the model of real knowledge in relation to which other branches are evaluated. Of course one cannot deny the existence of practical differences between the "two cultures"—the disciplines are not all the same or on the same footing; the point is rather that the present disciplinary framework is essentially underwritten by the rationality and conceptions of natural science and that this context has a decisive role in the attempt to establish the relative merits and demerits, successes and failures, of the different disciplines. Humanist theorists who reject what they vaguely refer to as positivism and who try to criticize the myth of an overarching scientific rationality are essentially asking that a different yardstick be applied to their discourses, yet it may be that the only yardstick of knowledge that can reasonably be applied is in a basic manner already that of science itself.

Yet the science in question here should not be understood in terms of the Laplacean myth, a myth that underwrites a sterile and unrealistic opposition between the hard and the soft, the objective and the subjective, the exact and the fuzzy, the useful and the superfluous, the simple and the complex. At the heart of such oppositions is the illusory image of a perfectly deterministic natural science that grasps the movements of a timeless nature in terms of a set of universal and perfectly necessary laws. In relation to this image, the world of creation and of history, the world of human action and of social life, is indeed irreducible to science, for its complexities and singularities are the work of processes we can describe only in terms of both necessity and chance, dynamics and history, global laws and local conditions. It is in the complex interplay of the two terms of such oppositions that whatever knowledge we have in fact arises. But what it is crucial to observe at this point, what has unfortunately been overlooked by many scientists and humanists alike, is the possibility that this opposition *of principle* between the exact and inexact sciences is largely a fiction based upon an unrealistic view of the former, and that a correct conception would be one in which the terms of this same opposition are reinscribed on both sides of the natural science–human science dichtomy. Such a possibility is at stake in Ilya Prigogine's comments that in contemporary physics we are beginning to understand that matter itself is to a certain extent a historical object, and that in the world of complexity—which is the real world of both matter and life—irreversible and stochastic process are an essential and ineliminable dimension, a limit not only to what can be known but to what is.[7]

[7]Ilya Prigogine, "Nouvelles perspectives sur la complexité," in S. Aida, P. M. Allen, H. Atlan, et al., *Science et pratique de la complexité: Actes du colloque de Montpellier* (Paris: La Documentation française, 1986), 129–43; "Order out of Chaos," in *Disorder and Order*, ed. P. Livingston, 41–60; "The Rediscovery of Time," *Zygon*, 19 (1984), 433–47; and "La Réconciliation d'Einstein et de Bergson," *CoEvolution*, 7 (1982), 28–31. For general background, see Erich Jantsch, *The Self-Organizing Universe: Scientific and Human Implications of the Emerging Paradigm of Evolution* (Oxford: Pergamon, 1980); and Arthur Peacocke, "Thermodynamics and Life," *Zygon*, 19 (1984), 395–432. For a clear presentation of the Belousov–Zhabotinsy experiments to which Prigogine refers, see Arthur T. Winfree, "Rotating Chemical Reactions," *Scientific American*, 230:6 (June 1974), 82–95.

CHAPTER 4

Idealism and Naturalism

What is implied in the very delimitation of a sphere that would be the object of the human sciences? What is presupposed in order to make the two conceptual steps whereby first the unity of the human (species?) is established and second the realities enclosed within this domain are distinguished from everything else? Are certain kinds of knowledge required in order to think in this manner? If so, this background would be common to all natural science–human science dualisms and could have some interesting implications for their status. Along these lines, one of Husserl's remarks is highly pertinent: "If I regard myself as a human being, I presuppose the validity of the world," he wrote in his 1931 lecture "Phenomenology and Anthropology."[1] And in such a context the notion of the "validity of the world" has far-reaching implications; for example, it implies that subjectivity and consciousness are attributes of human beings and not original and transcendental realities. This view amounts to an acceptance of what Husserl calls the "natural attitude," which places its faith in the existence and objectivity of a natural world prior to any constitution by the transcendental subject or "accomplishing subjectivity." For Husserl, this natural attitude, be it manifested in prescientific experience and activity or in the abstractive theoretical practices of the natural sciences, is essen-

[1]In Peter McCormick and Frederick A. Elliston, eds., *Husserl: Shorter Works* (Notre Dame, Ind.: University of Notre Dame Press, 1981), 315–23, citation: 319. This volume contains an extensive bibliography.

tially inadequate and provides the wrong starting point for knowledge. And in his view, this is so because the natural attitude conceals the active subjectivity that is the ultimate ground of all cognition. Only a philosophy that frees itself from the "realism" of the natural attitude and achieves a transcendental reflection upon the constituting ego, only a science that finds its ultimate and only starting point in the absolute purity of an "I think," can really lay the ground for an anthropology that does not grant in advance the validity of a naturalistic psychology, for which the conscious life is but an "annex" or "abstract stratum" of a corporeal "human" being existing within a natural and material world. In other words, Husserl recognizes that the psychophysical "dualisms," and with them the debates over the *Geistes-und Naturwissenschaften,* have an abortive character as long as they are conducted within the implicit framework of the world as it is given by the natural attitude and by the complex sciences that have proceeded from this attitude. As long as the debates occur in this framework, the methodological differences used to characterize the humanistic disciplines will be only local differences inscribed within the sphere of the natural sciences, sometimes in the form of positive results, but more frequently, it appears, in the form of lacunae, zones of silence and obscurity, or as an "irrational" desire to draw an arbitrary limit to the success of a project that has already demonstrated its remarkable capacities of discovery and explanation.

Husserl's Constitutive Subject

The thrust of Husserl's remark about the implications of speaking of a human sphere is to cast in doubt the cogency of any extreme dualist position that seeks to trace a limit to the domain of nature by circumscribing some particular sphere of "the human" inside this domain; for as long as the nature in question is the same nature that is conceived of by those sciences that arise from within the natural attitude, any limits inscribed within it will be in some important sense inessential. The "human" defined in this manner certainly may not be fully reducible to nature, but it remains a reality positioned *within* such a nature and subject to many of its determinations. Certain of its determinations may very well be different in important ways, but the basic principles of inquiry as well as important material conditions will not vary essentially across this sort of regional boundary. As far as Husserl is concerned, to speak of "the human" is to speak only of a mankind bound within nature—at least until the needed transcendental reflection has been achieved, at which point it is possible to speak of "concrete transcendental subjectivity" without taking the objective, natural world as the starting point and prior reality. Thus Husserl wants to replace all dualist conceptions with a monism that finally englobes the natural attitude and its objectivities within an overarching transcendentalism. From the perspective of his all-encompassing transcendental science that has its ultimate ground in a transcendental subjectivity, the nature studied by exact science is not even an anthropological constant but rather a local phenomenon, belonging to the surrounding world of only those who have learned how to adopt a certain attitude.

Although I am highly skeptical about the viability of Husserl's idealism as a response to the dangers of certain naturalist approaches to human culture and history, I do think that his radical perspective makes it possible to focus some of the basic issues quite sharply. His arguments suggest that it may indeed be difficult or impossible to hold a consistent, purely methodological form of a radical dualism. Deep-seated ontological assumptions are needed to underwrite a belief in the sheer inapplicability of the *fundamental* principles of scientific inquiry within some zone of the natural universe. Yet once this step has been taken, and as a result of a dualist ontology, the universe in question ceases, strictly speaking, to be the natural universe and becomes a universe where nature is paralleled by some other forms of being. Thus what is at stake are assumptions about an essential anthropological difference whereby mankind or "the human" exist in a manner that is not natural at all. Husserl's point seems to be that even this step cannot consequentially be taken unless we are willing to go all the way and grant the constitutive priority of the "I think."

What does this demand mean, and what are its implications for the attempt to establish a humanist discourse that does not have to borrow its footings from the very scientific conception that has already virtually written it out of existence? Although I cannot hope to do justice here to the many implications of Husserl's philosophy or, more generally, to the transcendental philosophical tradition of which he is but one part, I find it useful to contrast the minimal and local forms of dualism to Husserl's radical perspective. But first a few preliminary remarks on the place of Husserl and transcendental idealism within the contemporary intellectual landscape. For many, of course, this kind of literature represents a stage in the history of ideas that now seems passé; if the continuation of Husserl's project was once the goal of a large, international movement, today relatively few thinkers take an interest in phenomenology that is not purely historical or documentary. Roughly speaking, this minority can be divided into two groups. The first are those who work within the influence of a so-called analytic and empirical tradition whereby philosophical and sociological theorizing at once partake in, guide, and are guided by scientific research (philosophy as queen *and* handmaiden of the sciences, in a tangled hierarchy). Insofar as such thinkers take Husserl seriously at all, they are at pains to show how some of his insights can be useful to a "science" that bears little resemblance to the radical project of transcendental philosophy's overturning of the naturalist attitude.[2] Some of what Husserl says about intentionality, for example, could be adopted by the research programs of so-called cognitive science; moreover, some of his remarks on the constitution of meaning within the "life-world" may be usefully incorporated within the framework of an empirical and realist sociology. The second kind of contemporary interest in phenomenology is more frequently observed within more overtly literary and humanistic circles, where, oddly enough, Husserl and related figures serve as the representatives of a kind of tradition that is thought to be a major threat, badly in need of criticism. For example, Husserl, and behind him, Hegel and Kant, represent overly confident knowledge claims embodying a rationalistic hubris. Some of the criticisms and deconstructings of Husserl's positivities following from such a view seem poorly integrated within the larger philosophical and intellectual reality within which they occur; if what one is concerned with is the possibility and goal of a positive science of language, meaning, and interiority, an attack on Husserl's foundations, no matter how rigorous and devasting, is rather beside the point unless one has already agreed with Husserl's notion that the sciences cannot be discussed on their own grounds, since they have none. First one allows Husserl to place the natural sciences and human sciences as a whole within the ambit of the problematic of the transcendental subject, then one proceeds to a rigorous immanent critique of the failures and incoherence of his foundational project. The conclusion drawn from this exercise is that the scientific project has been given a thorough flogging! Husserl, then, who from the other side of the fence appeared as the last staunch defender of humanistic idealism,

[2]An example is the study by David Woodruff Smith and Ronald McIntyre, *Husserl and Intentionality: A Study of Mind, Meaning, and Language* (Dordrecht: Reidel, 1982).

is seen as the very embodiment of the hated scientific rationalism, and it is sufficient to subject him (along with Descartes, in passing) to a tough reading in order to take care of the menace.

In the present context, Husserl's philosophy is more accurately seen as a sustained attempt to limit the sphere and validity of the natural sciences without assuming the imposture of an incoherent dualism.[3] Yet insofar as Husserl's idealism is not finally freed of the dualist elements, this attempt must be judged a failure. One could, I think, write a fascinating book tracing the ontological equivocations involved in Husserl's various uses of the expressions *konstituieren* and *sich konstituieren*, for in spite of the many explicitly idealist passages in his work, Husserl is not always willing to state flatly that his "accomplishing subjectivity" is responsible for the creation or production of the very *existence* of the objects it intends. And yet the notion of the "constitution" of the object by consciousness often slides in this direction, so that if properly speaking the word is only used to signify the "becoming-present" of the object, that is, the object's mode of givenness to the subject, Husserl frequently inflects it with a more poietic sense. The equivocation involved in Husserl's usage of the concept is a slippage back and forth between a formal and a material sense of "constitute"—the former being a question of organization or arrangement (*einrichten*, *bilden*, etc.), the latter a matter of creation, making, and causation. These various senses of the term can be plotted along a scale in the following manner, marking off four typical stages:

1. At one extreme we situate the purely formal sense of "constitution" whereby no ontic claims whatsoever are made about the provenance of the object or the ontological status of the constituting ego, which is described as only the kind of "pure mental activity" that Kant made the only legitimate subject of "rational psychology." Here we can situate the famous example of the way physical objects are given to experience in a series of *Abschattungen* or perspectival "shadowings," never all at once. Another clear example is that of such phenomenal objectivities as rain-

[3]What's wrong with dualism? some readers may ask. As space does not permit the rehearsing of the many arguments, I limit myself to the following. If an immaterial, nonphysical mind or soul exercises a causal effect on a physical body, energy is created, which amounts to a violation of a principle central to physical science. But if mind or soul do not exercise any such effect (as in the epiphenomenal view), they amount to hypotheses having no explanatory role whatsoever in regard to human action, and there seems no ground for retaining them. For more detail, see Mario Bunge, *The Mind–Body Problem* (Oxford: Pergamon, 1980).

bows, the colors of which are constituted by the perceiving subject.

2. Next we find a usage of "constitution" which serves to designate the ways consciousness bestows meaning or *Sinn* upon the phenomenal data, synthesizing them in the positing of whole entities instead of discrete perceptual bits. Here we find the descriptions of the constitution of experience within certain organizational frameworks, such as spatiotemporal horizons or the figure-ground dichotomy.
3. "Constitution" as all of the above, yet also as the "spontaneity" of the subject: In this sense consciousness is constitutive in that it is activated by its own self-generating dynamic. Directing itself, it "holds sway" within its self-constituted field of experience and attention.
4. Finally, "constitution" as nothing less than "world-building." In this usage constitution is synonymous with productive causation or the making of external reality. All of the structures of intentionality, including reference to external objectivites, are purely autonomous and spontaneous, and consciousness becomes what Merleau-Ponty, in an idealist flight, called "le milieu universel."[4] In other words, this sense of "constitution" effaces the difference between real and purely intentional objects, between transcendent objects having autonomous being and those only "organized," "shaped," or "formed" by consciousness. An example of this fourth usage in Husserl can be found in sections 50 and 51 of *Ideas,* where he describes the entrance into the pure field of consciousness by means of the reduction or epoche:

> Instead of naively *carrying out* the acts proper to the nature-constituting consciousness with its transcendent theses and allowing ourselves to be led by motives that operate therein to still other transcendent theses, and so forth—we set all these theses "out of action," we take no part in them; we direct the glance of apprehension and theoretical inquiry to *pure consciousness in its own absolute Being*. It is this which remains over as the "phenomenological residuum" we were in quest of: remains over, we say, although we have "Suspended" the whole world with all things, living creatures, men, ourselves included. We have literally lost nothing, but have won the whole Absolute Being, which, properly understood, conceals in itself all transcendences, "constituting" them within itself. . . . it [this Absolute Being] is *essentially* independent of all Being of the type of a world or Nature, and it has no

[4]Maurice Merleau-Ponty, *La Structure du comportement* (Paris: Presses Universitaires de France, 1942), 144. The expression figures in the context of Merleau-Ponty's critique of the Gestalt psychologists' attachment to a realistic and naturalistic conception.

> need of these for its *existence*. The existence of what is natural *cannot* condition the existence of consciousness since it arises as the correlate of consciousness; it *is* only in so far as it constitutes itself within ordered organizations of consciousness.[5]

In this passage, Husserl's use of the notion of "constitution" moves squarely to the latter end of our scale of meanings, with only an inexplicable adoption of prophylactic quotation marks, a kind of bracketing within the brackets, remaining to attenuate the full-fledged idealism of a self-positing Spirit generating Nature within itself as one of its moments. Yet Husserl's previous remarks on causality nonetheless indicate an unwillingness to make his notion of "constitution" a matter of a quasi-divine reality creation pure and simple: consciousness is "self-contained" and "Absolute," but only in a sphere that is "outside" of all space and time; and if it is immune from the causal influence of nature and world, it is at the same "time" incapable of exerting causal influence upon anything else. Yet it seems that taking this last stipulation seriously makes it hard to understand in what sense we are to construe the statement that nature only *is* in so far as it "constitutes itself" within consciousness. It is true that the nature that existed before the evolution of sentient beings is an existence posited in a "transcendent thesis" of a consciousness. Or more simply put, it is true that the nature that existed before there were any minds is still a nature thought about by us. We can *know* its mode of being only in terms of the mode of being that we assign to it within our own cognition. But these latter clauses concern the order of knowing, and it is an error to grant consciousness or mind a similar priority within the order of being: it does not follow from the "indubitable" epistemic priority of mind (no knowledge without a subject) that nature can exist if and only if we happen to be able to assign to it the correct mode of being; it does not exist *because* we think about it or see it, it does not owe its existence to any of our cognitive acts of "constitution." Unless one holds to a radical constructivist account of natural science, unless, for example, one is prepared to posit the existence of a thinking substance possessing a fully autonomous existence independent of material conditions, it makes little sense to speak of the self-constitution of consciousness, of its Absolute Being independent of causal conditions outside itself. Kant was much more prudent as long as he

[5]Husserl, *Ideas: General Introduction to Pure Phenomenology*, trans. W. R. Boyce Gibson (New York: Collier, 1962), 140, 142 (Husserl's emphases).

sought to limit the scope of rational psychology's investigations of the indubitable "I think"; briefly put, the whole point of his "Paralogisms of Pure Reason" is as follows: although the concept of the *existence* of the "I" may be analytically contained in the *cogito*, nothing else follows—and no substantive features to be adduced in favor of either materialism or idealism can be derived from it.[6] Husserl, however, knows no such restriction, and he interprets the apparent necessity and indubitability of the "I" in all experience, finally, as proof that consciousness is essentially prior to its natural conditions and can be isolated as the unconditioned and Absolute being. Although the ultimately theological bases of the transcendental reflection are muted in Husserl's meditation, it should be recognized that what is being described here under the guise of a transcendental subject is a consciousness that has taken over the role played by the divinity within earlier systems. If Husserl has claimed that to speak of the "human" is to assume the validity of the natural world, we may add that to speak of the "transcendental subject" is to assume the validity of a supernatural world, that of a disembodied and self-engendering consciousness. Indeed, many of Husserl's formulations strongly suggest that he thought that this milieu of pure thought was the one true world, and that the "natural" world, the world of human, "incarnate" subjects, ultimately exists only within this ideal sphere.[7] His philosophy, then, is that of an idealistic monism, as is indicated by his statement that natural science should be understood as one spiritual formation among others, and that its object—nature—exists only for the initiates. In this manner Husserl charges the humanistic knowledges, in his Vienna lecture of 1935, with the task of becoming the ultimate metalanguage, capable of englobing their opponent—the natural sciences:

> What is obviously also completely forgotten is that natural science (like all science generally) is a title for spiritual accomplishments, namely, those of the natural scientists working together; as such they belong, after all, like all spiritual occurrences, to the region of what is to be explained by humanistic disciplines. Now is it not absurd and circular to want to explain the historical event "natural science" in a natural-scientific way, to explain it by bringing in natural science and its

[6]Kant, *Critique of Pure Reason*, trans. Norman Kemp Smith (New York: St. Martin's, 1969), A 339–405; B 397–432. All subsequent references are to this edition.

[7]A reading of Husserl that supports this view is that of Paul Ricoeur, "Phenomenology and Hermeneutics," *Noûs*, 9 (1975), 85–102.

> natural laws, which, as spiritual accomplishment, themselves belong to the problem?[8]

In this passage Husserl explictly asserts his belief that the natural sciences cannot possibly inform or provide their own philosophical self-understanding or theoretical metalanguage; given this premise, it falls upon his own philosophy, a philosophy of *Geist*, to stand above the other, local knowledges as the overarching source of reflection. Once replaced within this properly spiritual view of human history and thought, the sciences recover their proper role and can potentially regain the meaningfulness that was forgotten as scientific research become a matter of an abstract technique detached from its sources in the "life world." Thus the art of engineering, which supplanted the philosophical attitude of *theoria*, should be returned to its proper place. Yet the cost of this recovery of science's meaning seems to be that implied in Husserl's unflinching use of the term *Geist*: idealism.

The Slide to Dualism

Although it is perhaps possible to conceive of a coherent form of idealism, what is difficult to imagine is a coherent idealism that incorporates within itself many of the descriptions of reality arising from within a naturalist or dualist perspective. And this is precisely what we are given when Husserl begins to speak of a "concrete transcendental subjectivity," by which he means to refer to the empirical subject or human-being-in-the-world. When all is said and done, this subjectivity's knowledge of the world and action within it obey many of the same constraints that a naturalist conception insists upon; for example, this subject is situated in time and space, cannot engage in "magical" forms of causality, is mortal, eats and drinks, participates in social life. Of course there are those who believe in an unmitigated idealism or spiritualism, who maintain that matter is really spirit and that some empirical subjects enjoy an omnipotence of thought, immortality, and so on. But insofar as the methods of experimental scientific investigation have been unable

[8]Husserl, "Philosophy in the Crisis of European Mankind," in *The Crisis of European Sciences and Transcendental Philosophy: An Introduction to Phenomenological Philosophy*, trans. David Carr (Evanston, Ill.: Northwestern University Press, 1970), 272–99, citation: 272–73.

to provide any reasonable warrant for such fantastic hypotheses (and indeed go very far toward totally destroying their credibility), no plausible idealism can today take this exceptional path. Phenomenology, insofar as it lays claim to being a form of serious inquiry—and indeed, a science—surely must abstain from making hypotheses that overtly contradict the findings of the natural sciences and hence must lend no credence to supernatural and mythical entities. At the stage in history at which Husserl is writing, philosophy and the discourses of the humanities in general cannot simply turn their backs on the results of the natural sciences, they cannot simply arrogate to themselves the autonomy required to set up, in isolation, their own criteria of knowledge, truth, and reality, building for themselves their own idiosyncratic universes. Or rather, we should say that insofar as a thinker simply ignores the natural sciences in this manner, he or she seriously limits the scope and plausibility of the theses defended and runs the risk of the most flagrant impostures (that, for example, of the twentieth-century *Lebensphilosoph* who, when it comes to his own ailments, suddenly places the greatest confidence in a medical science that flatly denies both the theses and the methods of a romantic conception of living beings; or again, the critic who promulgates a totally romanticist view of nature and who nonetheless goes about disseminating this vision by means of typewriters, computers, printing presses, industrial systems of transportation—none of which could be plausibly explained within that vision). It is to Husserl's credit that he does not countenance such a stance, for although he cannot accept the modern scientific worldview, he recognizes the necessity of directly confronting a reality that has acquired a cultural and historical significance too great to be ignored. Phenomenology must englobe, and not simply find a place alongside, the natural sciences, for failing this, there remains the possibility that one day the sciences will finally succeed in offering a better explanation of phenomenology than the one phenomenology provides of itself. But to englobe the natural sciences is not to refute or deny any of their specific or positive theses by reference to norms of coherence or additional evidence—for it is precisely the business of science to be engaged in that sort of activity on a large-scale, ongoing basis. *That* sort of refutation is provided by science itself, which only provides additional confirmation of the critical and progressive nature of the scientific project. In this sense Husserl's disagreement with science can be situated only at certain levels, the foundational ones, and therefore must leave others largely, if not wholly, intact.

But does not an idealism cease in this manner to hold firm to its positing of a purely spiritual universe? It begins, for example, to adopt a language of "concretization," "embodiment," and "incarnation," and its "subject" ceases to be an angel, demiurge, or self-positing spirit, being described instead as a natural and social entity, one who comes into being and passes away within the "life world"—that halfway house between an ephemeral universe of pure idealities and a physical nature. Consequently, we are warranted to ask whether Husserl's philosophy does not also contain within itself a mysterious form of dualism, which, far from providing the promised resolution of the dualisms supposedly caused by the realist and natural scientific attitudes, merely reiterates them in a more cryptic form. What is especially mysterious and cryptic in this kind of dualism is the general rationale for the mixings and contacts between two heterogeneous "substances" or realities, and most of all, the reason for the embodiment or incarnation of such "immaterial" or "nonphysical" entities as "the subject," "ideas," or "meaning." Why does "the subject," if it is pure spontaneity, self-constitution, and activity, get mixed up in the kinds of processes that we associate with the word "life"—such as *biological* sexual difference, aging, sleep, and eating, to name only the most obvious, minimal necessary conditions of human existence? If meanings are spiritual or wholly subjective, why and how is it that they have the habit of embodying themselves in such things as ink and paper, or in hardware? Husserl's complaint about Kant is that, because he began with a naturalistic psychology, he could not elevate the subject from there to its properly transcendental status.[9] But Husserl himself grapples with the converse failing, which is but the other side of the same dualist coin; having purified the subject to the point of "Absolute Being," he has difficulty explaining its manifestation in the form of living creatures. Why is the "life world" that is constituted by concrete subjectivity a world of "life" and not something else entirely? It seems that to conceive of a subject that is "human" is to conceive of one whose existence is conditioned, in part, by certain biological processes that, although they have never been fully explained, have nonetheless been the object of remarkable discoveries within the life sciences—discoveries, such as those of genetics and molecular biology, that simply are not *based* upon the inner phenomenologies of living processes.

Here we touch upon an interesting chapter in the intellectual

[9]Husserl, *Crisis*, §§ 30–31.

history of dualist arguments. As long as vitalist, animist, and organicist views on the nature of living creatures had not yet been seriously rivaled by mechanistic and materialist conceptions, the realm of the living as a whole could be plausibly held to be inimical to certain kinds of explanations that were seen to be successful in regard to physical nature. With the advances of a naturalist and experimental life science, it became increasingly necessary to retrace the boundary, to draw the limit to materialist explanation not at life but at the specifically human (language, culture, etc.). In this regard, Kant is a particularly interesting figure, for although he wanted to maintain the special status of the living—as a self-organizing and telic unity—he also sought to reconcile these features with his causal and determinist categories of natural phenomena, the result being a conception that anticipates some contemporary discussions within theoretical biology. The line of resistence traced by Kant, however, seems to have been made at the cost of a large sacrifice: if living organisms are not to be explained in terms of mechanical causation, if a teleological judgment is made whereby the "causality of reason" is brought into play, this is to some extent only an epistemic thesis; it is a matter of an Idea and not a full-fledged ontological assertion; thus, whether the "purposive" causality of the *Naturzweck* is ultimately identical to a physico-mechanical process is held to be undecided.[10]

My cursory discussion of Husserl has been intended to bring forth some of the implications of the metalanguage that must be relied upon in discussions of the relationship between the natural and human sciences, particularly those discussions that seek to establish radical dualisms or that try to subsume the natural sciences within some broad humanist or idealist conception. My conclusion is that it is important to agree with Husserl that it is incoherent to conceive of a science *of humanity* having an object domain and epistemic principles fundamentally different from those of natural science, because the only consequential manner to break fully with the latter is to engage in a thoroughgoing romanticism or transcendental idealism whereby it is not only "the human" that soars above the sphere of

[10]The text in question is of course the second half of the Third Critique, "Critique of the Teleological Judgement," in *Critique of Judgement*, trans. J. H. Bernard (New York: Hafner, 1951), 205–339. Subsequent references are to this edition. For a valuable commentary, see Clark Zumbach, *The Transcendent Science: Kant's Conception of Biological Methodology* (The Hague: Martinus Nihjoff, 1984); one may also consult Claude Piché, "Les Fictions de la raison pure," *Philosophiques: Revue de la Société de Philosophie du Québec*, 13 (1986), 291–304. The contemporary discussions alluded to can be approached in Milan Zeleny, ed., *Autopoiesis: A Theory of Living Organization* (New York: North Holland, 1981), esp. Humberto R. Maturana, "Autopoiesis," 21–33.

nature to constitute its own autonomous world but "life" or "spirit" as a whole. Such is, it strikes me, the true heart of the phenomenological movement, and in this regard one can see that a great deal of the work done within the humanities today is still underwritten by similar assumptions; the history of human society and cultures is approached as so many different individual and collective spiritual expressions which, freed from material and natural constraints, coexist as part of an unlimited domain of meaning. The "texts" that are the windows to this domain are to be compared and contrasted, interpreted and understood in an unlimited and free series of "communications." For Husserl, this process of *Besinnung* was meant to recover the overall goal and order of the development of thought; today it is often directed only by the putative spontaneity of the individual interpreter's curiosity or pleasure. Here we are indeed confronted with one half of the classical opposition to which I referred earlier: culture is basically conceived as a detached and autonomous realm of idealities, of meanings and visions that bear absolutely no necessary relations to any nonidealistic conditions. Following this basic premise, it is the task of the natural sciences to deal with the realm of necessity, where it is possible to trace an event back to its governing conditions; the humanist, on the other hand, is concerned with the free play of a separate domain of expressions, interpretations, and values.

Yet it should also be noted that the humanistic readings and interpretings, be they critical or sympathetic, deconstructive or hermeneutic, evaluative or descriptive, have nothing to do with the raison d'être of Husserl's phenomenological project, which they resemble as little as the pile of sherds resembles the vase. This is so first because a generalized skepticism has vitiated the vigorous sense of inquiry that motivated Husserl, and second because an isolationist and fragmentary approach has replaced Husserl's urgent sense of the need to ascertain the place of his reflections in relation to the sciences and the larger problems of knowledge they bring to the fore.[11]

[11] The same admirable feature of phenomenology, as well as the same failures, are manifested in Maurice Merleau-Ponty's confrontation with the behavioral sciences of his time in *La Structure du comportement*, a work that embodies all the problems we have just discussed in relation to Husserl. Here as well the desire to surmount metaphysical dualism leads to an idealism that embraces matter, life, and spirit within itself as "orders of meaning." The result, however, is a certain incongruity, for we wonder why the findings of experimental and Gestalt psychology should matter at all to a conception of cognition and behavior in which it is taken for granted that "perception is not a natural event" (157, cf. 207, 208, 214–15) but is, on the contrary, "constitutive" (235–36) in Husserl's transcendental sense.

In this sense—but only in this sense—certain contemporary romanticists may be right in throwing Husserl into the same sack as their accursed enemy, science, for although Husserl was opposed to the claims of the naturalist attitude, he was critical of the various facile ways in which philosophy could overturn this rival.

The Limits of Naturalism

Husserl's project highlights the importance of the starting point and metalanguage that make possible any local or specific discussion of the relations between the natural sciences and a knowledge of humanity. Bluntly put, the position being forwarded here is that, if it still made sense in Kant's time to hold to the necessity of radical differences between our knowledge of physical realities or nature, on the one hand, and living beings, on the other, such radical dualisms find little support in the achievements of the biological sciences to date.[12] If it is impossible to withdraw the "mystery of life" from the ambit of science as a matter of principle (or fact), and if it is granted that human beings are also in some important sense living creatures, we are obliged to recognize the reality of a scientific knowledge, however partial, pertaining to human beings. Nor does this knowledge address only the physical attributes and rudimentary biological bases of human existence; I am not leading humanists up to the stunning conclusion that the study of the demographics of human populations, nutrition, and similar topics is best conducted in the light of contemporary life science. Insofar as such capacities as perception, cognition, and intelligence are thought to be part of what is essentially human—or at the very least necessary (but not sufficient) for its full development—the natural sciences do indeed have a hold on the subject; for it is only within their framework that the origins of sentience receive anything approaching a coherent and well-based explanation.

In order to suggest how this latter assertion could be supported, one need only evoke evolutionary epistemology, a large and well-developed research field.[13] Generally, the evolutionary account of

[12]For background on the biology of Kant's time, see Thomas S. Hall, *Ideas of Life and Matter*, 2 vols. (Chicago: University of Chicago Press, 1969), 2:5–118; and Elizabeth B. Gasking, *Investigations into Generation, 1651–1828* (Baltimore: Johns Hopkins University Press, 1966). For a more recent conception, see Jacques Monod, *Le Hasard et la nécessité* (Paris: Seuil, 1970).

[13]See Gerhard Vollmer, "Mesocosm and Objective Knowledge: On Problems Solved

the "miracle" of human intelligence begins with the thesis that, if the brain possesses the astounding capacity of having certain of its states link up with the realities of its environment, this is only the natural (albeit highly fortuitous) result of the evolutionary history that caused the whole organism to come into a viable form of existence in the first place. This view of the emergence of sentience has been fleshed out with some detail and finds its bases in current theorizing and experimentation concerning neurophysiological and neural organization; it receives further support from evidence provided by the fossil record and from studies of living organisms having a neural organization that is thought to be primitive enough to offer the possibility of fruitful analogies. The genesis of sensory and central nervous systems in living organisms is said to have become possible through the emergence and specialization of cells that are capable of communicating electrochemical impulses within the organism when there is a disturbance on the cell's membrane; these impulses can be triggered by different kinds of disturbances or environmental perturbations, such as changes in physical pressure, temperature, or illumination. In time, the nerve cells become more and more differentiated and complex; through genetic mutation, for example, a light-sensitive surface on an organism begins to provide rough directional information about light sources (and indirectly about movements in the immediate environment), and such information directs adaptive motor responses. Eventually there emerge organisms having complex systems of sentience enabling them to respond to pertinent features in the environment with greater and greater degrees of accuracy, and finally it becomes possible for certain internal states of the organism to *refer* to aspects of the world.[14]

It is a basic tenet of evolutionary epistemology that it is possible to offer a fully naturalistic account of the origins of intelligence. Although many of the pieces in this puzzle are presently missing, the general approach seems correct. At the present time, it offers the most detailed, coherent, and well-supported hypotheses as to the origins of cognitive capacities, and thus it stands as a formidable rival to idealist ontologies in which thought is held to be free-floating and absolutely self-creating. Yet in the context of the status of humanistic knowledge, a large question remains open, a question about what

---

by Evolutionary Epistemology," in *Concepts and Approaches in Evolutionary Epistemology*, ed. Frank M. Wuketits (Dordrecht: Reidel, 1984), 69–121; and *Evolutionäre Erkenntnistheorie* (Stuttgart: Hirzel, 1981).

[14]Paul Churchland, *Matter and Consciousness: A Contemporary Introduction to the Philosophy of Mind* (Cambridge, Mass.: M.I.T. Press, 1984), 121–29.

follows from the acceptance of a naturalistic starting point and perspective. It has been claimed frequently enough that everything that is of interest to the humanist in fact remains untouched, so it is necessary to address in more detail the objection that the scientific study of man is incapable of seizing "the essentially human" following any of its important or interesting definitions. In regard to the specific case of evolutionary epistemology, such reservations appear to be well founded. Gerhard Vollmer, for example, does not believe that a confidence in evolutionary epistemology's explanation of the origins of cognition entails a belief in the possibility of fully explaining human knowledge in the same manner. He notes that evolutionary epistemology does not preclude the possibility of *cultural* evolution and history. In fact, it is important to distinguish between mankind's "cognitive abilities," on the one hand, and "human knowledge," on the other, for the latter has a more restricted sense. Knowledge, it should be recognized, is biologically conditioned, but it is not completely determined by these conditions. To hold that human cognition arose as a result of wholly natural and material processes does not require one to think that an explanation of its emergence also provides us with a solution to all epistemological problems—such as a theory of truth.[15] The naturalized epistemology that points to the biological bases of "successful" cognition is a crucial part of the story of human knowledge, but it cannot provide a full account of the complex history of human beliefs, of their truths and of their errors. In a sense, a full-fledged naturalistic epistemology can only lead to the threshold of that problem.[16] Even once it is granted that a certain kind of biologically evolved cognitive system is a necessary condition for human capacities, there seems to remain an important limit to the implications of this truth. Thus it seems unlikely that a more precise knowledge of human genetics will specify anything like a closed definition of what is essentially human: on the contrary, it is more probable that such research will provide additional confirmation of present hypotheses that the particular organization of human brains is largely conditioned by the history of their own functioning, that the brain is thus a self-organizing system, made possible, but not overdetermined, by its genetic bases and

[15]Vollmer, "Mesocosm and Objective Knowledge," 85.

[16]Thus I diverge at a basic level from the kind of defense of scientific realism represented by Michael Devitt in *Realism and Truth* (Princeton: Princeton University Press, 1984).

[17]Michel Morange, "Biologie moléculaire et anthropologie," *L'Homme*, 26 (1986), 125–36. On the concept of self-organization and recent models in cognitive science,

environment.[17] Such considerations undermine both the plausibility and danger of the kind of naturalistic reductions dreamt of by nineteenth-century naturalists and their sociobiological avatars. Clearly that kind of theory cannot give a detailed account of human knowledge. How could it ever explain in solely biological terms the fact that the element germanium was not discovered until 1885? The history of the emergence of this kind of scientific inquiry must uncover the social conditions that made it possible.

But to discover limits to the implications of a naturalist perspective is not to proclaim in advance that there are *no* implications for research in the human sciences, for it should be clear that even the naturalist's minimal account of the basic conditions of possibility of human thought, belief, and emotion render implausible a very large number of mankind's past representations of his own origin and place in the cosmos—supernatural and anthropocentric conceptions, for example, as well as those in which the most banal and well-established natural regularities are violated. The implications that I have in mind are epistemological as well as ontological. The former amount to a recognition of a basic continuity between the fundamental principles of research and inquiry within the natural and human sciences, a unity that does not amount to a requirement that precisely the same specific procedures be applied in all cases, or even to the more general view that there is only one acceptable type of explanation, specifiable in logical form. The other implications generally concern a limitation on the kinds of entities and processes we are warranted to acknowledge as being at work in human history.

These two topics are taken up in greater detail in the subsequent chapters. Yet I must clarify right away what is meant by the ominous-sounding notion of the "ontological implications" of natural science for research in the humanities—it being stressed in advance that the last thing I have in mind is the goal of establishing a set of a priori theses about the Being of Man and of Society from which one could infer the guiding principles of research. Nor do I advocate a revamping of the evolutionary and biological approach to culture attempted by Hippolyte Taine.[18] Such impulses run wholly contrary

see Paul Dumouchel and Jean-Pierre Dupuy, eds., *L'Auto-organisation: De la physique au politique* (Paris: Seuil, 1983); and D. E. Rumelhart and J. L. McClelland, *Parallel Distributed Processing*, 2 vols. (Cambridge, Mass.: M.I.T. Press, 1986); Paul Smolensky, "Formal Modeling of Subsymbolic Processes: An Introduction to Harmony Theory," in *Advances in Cognitive Science*, 1, ed. N. E. Sharkey (New York: John Wiley, 1986), 204–35.

[18]See, for example, *Philosophie de l'art* (Paris: Fayard, 1985), a work of 1865 in which the emphasis on a process of "selection" is made clear (46).

to the epistemological norms that must guide nondogmatic, open-ended, critical inquiries. And these norms are indeed the first and most important resource that the scientific model of knowledge has to offer the humanist.

Someone who believes in an idealist metaphysics can feel perfectly justified in doing intellectual, and more specifically, literary, history in a way that simply does not go without saying once materialist objections are raised. The idealist believes that he or she is well founded, for example, in tracing a disembodied ideal or aesthetic entity across historical and cultural boundaries, describing the differences and similarities of its different manifestations. This kind of work proceeds without it ever being shown what conditions (physical, biological, psychological, social) could have made possible the concept's real manifestation, a manifestation that should be recognized as relying upon the cognitive acts of some living agent existing within time and space. No account at all is given of why we should believe in the real existence of the invariant entity postulated in such descriptions. Once such a belief in the autonomous and independent existence of ideal concepts and of ideal aesthetic objects is put in question, the first consequence is a recognition of the necessity of addressing the ontological assumptions implicit in the interpretation of cultural processes, events, and artifacts. Some such assumptions are simply untenable, and to recognize this reveals the frailty of all of the results based on them. There is no reasonable warrant to believe in the workings of disembodied spirits and powers, yet they are implicitly taken for granted in any number of otherwise careful humanistic descriptions and analyses.

Ingarden and Idealism

I shall now try to illustrate the pertinence of my critique of idealism by making a few remarks on the aesthetic doctrine of Roman Ingarden—a doctrine that in spite of its various particularities amounts at base to a sustained effort to defend an idealist and organicist conception whereby the work of art, as the expression and embodiment of an Idea, possesses a perfect essence and self-identity. Although it is true that Ingarden thought his investigations into the ontology of works of art were a first step in the direction away from Husserl's transcendental idealism, from a materialist perspective this step is all too short. Ingarden "saves the text," that is, he argues for the literary artwork's essence and identity, at the cost of assuming in advance that there exists a closed and immutable realm of ideal concepts. Grant me the Platonic *eide*, and I will be able to save the one true *Hamlet* from the vicissitudes of time, space, and history.

Some of the details of Ingarden's arguments have important implications for the epistemology of literary criticism's objects and methods. My objections to Ingarden's "solutions" in no way diminish my sense of the value of his discussions of many of the issues; Ingarden's two books on the subject present rigorous discussions that have been repeated in a watered-down form in a stream of critical articles and books.[19]

Ingarden's name is most frequently associated with the theory of the "stratified nature" of the literary work—as it was vulgarized by René Wellek in *Theory of Literature.* He is also named as the grandfather of a "reception theory" of literary criticism which corrects the excesses of an author-centered approach. Neither of these labels has much to do with the thrust of Ingarden's two major books on literature.

Although Wellek speaks disapprovingly of the dangers of organic metaphors, he hardly seems aware of the various assumptions Ingarden relies upon in developing a theory of the identity and status of the literary work. This is a serious oversight because what is at stake are the very foundations of the distinction between intrinsic and extrinsic topics in literary analysis, a distinction that is essential to Wellek and Austin Warren's whole project of establishing the priorities for a coherent and autonomous literary discipline. A closer look reveals that the conceptual bases are anything but specific to the literary field.

What, then, are these overlooked issues and assumptions in Ingarden, and where does their analysis lead? We can deal with this question by starting with Ingarden's analogy between organic beings and literary works and then exploring the properly aesthetic assumptions that are conjoined with the organicist principles. What is left

[19]For what follows, Roman Ingarden, *The Literary Work of Art*, trans. Grabowicz (Evanston, Ill.: Northwestern University Press, 1973); all references to this edition cited as *Literary Work* with reference to section divisions followed by page numbers; *The Cognition of the Literary Work of Art*, trans. Crowley and Olson (Evanston, Ill.: Northwestern University Press, 1973), cited henceforth as *Cognition*, with references to section and page numbers; "Des différentes conceptions de la vérité dans l'oeuvre d'art," *Revue d'esthétique*, 2 (1950), 162–80; "A Marginal Commentary on Aristotle's Poetics," *Journal of Aesthetics and Art Criticism*, 20 (1960–61), 163–73, 273–85. Among the wealth of commentaries, see Yushiro Takei, "The Literary Work and Its Concretization in Roman Ingarden's Aesthetics," *Analecta Husserliana*, 17 (1984), 285–307.

when such assumptions are rejected is the object of the final part of this book.

Ingarden begins with a conception of the organism which is largely influenced by the work of Ludwig von Bertalanffy, the renowned systems theorist. An organic mode of organization is seen as having three primary characteristics: (1) the parts of the entity are functionally related within a whole, whereby the "meaning" of the organization resides in its fulfillment of a primary function—the maintaining of its own organization; (2) this functional mode of organization is realized through an ensemble of strong internal hierarchical connections; no part or organ of the larger system is self-sufficient, but the reciprocal interaction and cooperation of the parts amounts to the self-regulation of the whole; (3) the organism is characterized by a typical chronological development or life process, including, for example, stages of maturation as well as an eventual dissolution.[20] Having presented these points, Ingarden traces a limit to his analogy and asserts that a work of art *is not* to be viewed literally as an organism in this sense: "It is not, of course, a living organism and also has no ontically autonomous being. It owes its existence and its form to the creative acts of consciousness and other acts of the author."[21] In this regard, Ingarden echoes Kant, who pointed out that, although the parts of a human artifact such as a timepiece may indeed be reciprocally related within a whole, this relation is not one of *production*, as is the case of a *Naturzweck* or "self-organizing" being. The work lacks this "formative power" within itself; it is produced, indeed "finished" from without.[22] To view a work of art as an artifact having a source of efficient causation outside itself in its maker—the artist—is precisely *not* to view it as an organic or self-organizing being in the Kantian sense; in Ingarden, claims for the autonomy of the artwork find their ultimate limit in the necessity of a creating agent. Thus, when Ingarden goes on to specify how a work is like an organism, he retains only two factors: (1) the parts within the work of art (e.g., its various strata), while heterogeneous, are essentially interdependent and fit together precisely as parts within an ordered whole; and (2) this organization is such that the work is designed to exercise a "particular main function proper only to itself."[23]

[20]Ingarden, *Cognition*, § 13a, 72–90.
[21]Ingarden, *Cognition*, § 13a, 76.
[22]Kant, *Critique of Judgement*, §§ 65–66, 218–24.
[23]Ingarden, *Cognition*, § 13a, 78. The earlier work, however, includes a lengthy and very dubious extrapolation about the "life cycle" of the work of art; see *Literary Work*, § 64, 343–55.

Ingarden's specification of his notion of the essential function of the work of art, as well as his views on how this function is achieved, lead him even further from the metaphorical grounds that a theoretical biology could lend to an aesthetics (and then to a literary theory). Instead, he moves toward a more properly metaphysical statement of his assumptions about the nature of aesthetic value and meaning. In an important sense, a sense anticipated by Kant, it is already metaphorical to speak of the pure autonomy of a living organism, which cannot fully detach itself from the causal nexus of the physical environment within which it emerges—if only because, as an open system, it must exchange matter or energy with the outside to sustain its "autopoietic" processes. It would be yet another metaphor, a figure of a figure, to view an artwork as such an entity; the artwork is a designed artifact possessing only a "heteronomous" mode of existence, as Ingarden rightly puts it. But what is idealist about that? Is this not indeed the path *away* from idealism, and toward a view in which the work is perceived in terms of its necessary connection to external conditions? Yes and no, for when we turn to Ingarden's descriptions of the external bases of the work of art's existence, we find a real ambivalence, indeed, a fundamental dualism, at the heart of his conception. And this dualism, as may be expected, generates a series of inescapable confusions.

As has already been clearly indicated, Ingarden states that one of the external sources of the work of art is its maker. Consequently, it is the intention of the maker that endows the artifact with its function. So far, the picture is traced wholly within the schema of a teleological or functional mode of explanation based on the model of "design"—an explanation that is properly etiological and involves no ontological scandals. The artist is the efficient cause of the work, which can in part be explained and understood by reference to this agent's intent and action. Here we may introduce another major aspect of Ingarden's thought, his consistent struggle against what he knew as "psychologism" and the relativism to which it would lead. Psychologism is any conception that makes the ultimate reality of the work of art depend on the whims of the perceiver. An obvious example is the "reception theory" in which there are as many *Hamlets* as there are performances and readings. Thus there is a distinction crucial to Ingarden between, on the one hand, the real work and varied manifestations *of it*, and on the other, an open-ended range of misreadings that, though they take the work as a kind of catalyst or stimulus, do not amount to legitimate aesthetic responses to the work of art *in itself*. In Ingarden's jargon, the work of art, intended and created by the artist, is a "purely intentional object"—which amounts to saying that in order for it to be at all, it must be thought or conceived by a human intelligence. The work of art as read/perceived/concretized is what he calls a "derived purely intentional object"—which amounts to saying that although the perceiver is necessary for there to be aesthetic experience, *what* this perceiver experiences, the object, is derived from another intentional act, that of the creative moment. The reader, then, does not create the work—for that would only lead to the chaos and irrationalism of psychologism, a topsy-turvy world where, if the reader chooses, *Alice in Wonderland* can be recreated and taught as *(Ph)alice in Wonderland*, a fantastic erotic allegory about penis envy. (I include this example of creative criticism, which is of course not Ingarden's, to give the reader a hint as to why one might object to the seemingly liberating idea that the reader creates the work. I cannot, however, claim it as my own invention.)

Ingarden appears to be arguing solely for a fixing of the work's identity on the basis of what is now frequently referred to as "author's meaning." Such a view is frequently assigned to him. Yet this interpretation of Ingarden is beside the point. Not only are there hesitations on the question of authorial intention in Ingarden's text, but more important, to emphasize this question is to miss the real thrust of his aesthetics. To take up the latter issue first, Ingarden's conception of the work of art is one in which the work's meaning is not to be confounded with a set of notions that can be extracted from the work and stated explicitly in a process of depth interpretation. Ingarden does not inhabit a landscape scarred by the pollution left by the interpretive industries. The overarching function of the work is to express an "idea," but this idea is not reducible to a set of propositions which it is a critic's task to elucidate and publish. The "concretization" of the idea involves the experiencing of what Ingarden refers to as "metaphysical qualities," which, once more, transcend the art of hermeneutics.[24] Thus any debate over the problem of how the interpreter can "get the right meaning" is only remotely related to Ingarden's discussions of the relation between the work's schemata and its range of legitimate concretizations, for Ingarden is simply not talking about the problem of correctly disambiguating metaphorical and suggestive textual features. Indeed, he asserts that the literal translation of metaphor only produces absurdities and

[24]See especially *Cognition*, § 13, 79–90.

lies.[25] That kind of interpretive approach does not even figure in his hierarchical list of types of cognitive attitudes to literary works, for, as it is not even pertinent to the aesthetic experience, it is deemed wholly inappropriate.[26] Only once these points are grasped can we move on to the issue of author's intentions in Ingarden's framework. In several places, Ingarden speaks of one of the major ontic bases of the work as being the *conscious acts* of the author.[27] But he also argues very clearly, and in a crucial context—that of his direct attack on relativism—that to state the problem as a search for the author's thoughts at the moment of creation is an error that only prepares the way for a skeptical refutation based on the impossibilty of ever reconstructing the pristine moment of an originary intention, if by this is meant the immediate and full presence of a particular conscious act.[28] In order to understand Ingarden on anything approaching his own terms, one must be able to grasp how it is possible to assert the necessity of recovering a work's original meaning, a meaning directly related to the creator's intentions, *without* this entailing a psychological approach to the work of art. The key to this way of thinking is deceptively simple: if the emphasis on the creator's intentions works as a counterbalance to the vagaries of the readers' responses, this does not mean that the subjectivity of the artist is thought to be free, spontaneous, and autonomous in its creative, intentional acts. On the contrary, there is a limit to this creativity, and the ultimate bases of both creation and reception stand outside of subjective processes entirely (just as aesthetic value and the nature of the aesthetic experience stand outside history). Ingarden's understanding of this basis is pointed to in a crucial phrase: "Now everything depends on how it is possible to have an intersubjective common (*koinē*) language."[29]

What is it that makes such a language possible? How is any language at all possible? For Ingarden, the two questions are the same, for language begins with the possibility of an intersubjectively shared meaning, the possibility of thinking the same thought at different times, both of which possibilities require that different sounds, "acoustic images," and graphic inscriptions be recognizable as tokens of the same meaning-content. He implicitly reasons in the

[25]*Cognition*, § 13, 67.
[26]*Cognition*, § 25, 223.
[27]*Cognition*, § 31, 335.
[28]*Cognition*, § 31, 347.
[29]*Cognition*, § 31, 349.

following way: if there can be such a thing as a proposition (an identity of meaning across different sentences), then there can be such a thing as a work of art possessing an identity and essence not reducible to a particular physical object (or set of objects). Not to recognize such a possibility is, in his view, the end of thinking, so the only serious question concerns how this possibility of our cognition works. His answer is that ideal concepts and essences, such as "redness," have a real, immaterial, nonspatial, and atemporal existence. They neither come into being nor pass away, nor can they be created.[30] They serve as "regulative principles" whereby actual subjectivities "derive" meanings from physical inscriptions, sounds, and so forth, associated with these idealities by convention. So the subjectivities of the reader and author can neither individually nor conjointly suffice to explain the possibility of shared meanings and identical aesthetic experiences and judgments; it is true that for a sentence or work of art to exist, there must be subjects, but there must also be physical entities or vehicles ("signifiers" in another jargon) as well as ideal concepts: "For without ideal essences and ideas, purely intentional objectivities are impossible in the same degree as *real* objects in a *true* sense."[31] In this sense, the creator too only "actualizes" meaning possibilities that exist before the act of creation. Ingarden clearly affirms that the idealities are the basis of both the genesis or creation of the work as well as its continued existence after the act of creation is completed; this ideal basis is transcendent in relation to the work and stands as an *eidos* upon which the creation is modeled. It is not only the reader's creativity that is limited in this model but that of the author as well, who in a sense cannot be a real creator: "The intentional act of pure consciousness is not creative in the sense that it can create genuine realizations of ideal essences or ideal concepts in an object that is intentionally produced by it."[32] This last observation casts Ingarden's objection to Husserl's transcendental idealism in a different light, for the objection concerns the concrete subject's putative capacity to constitute a truly transcendent object; if Ingarden joins a materialist view in denying subjectivity the ability to create physical objectivities, he limits it in another dimension as well, for the subject can only discover, not invent, essential concepts. But from another perspective this is a pure vestige of idealism, for it amounts to

[30]*Literary Work*, § 3, 10.
[31]*Literary Work*, § 66, 363.
[32]*Literary Work*, § 66, 362.

postulating a dual universe, one where the literary work of art is neither real nor ideal but partakes of both kinds of substances in a mysterious combination.[33]

As I said at the outset of this discussion, the theory under consideration is question begging and gives itself in advance what it sets out to prove; to take for granted the existence of a knowable realm of eternal idealities is to evacuate the historical and social dimensions of human cognition, language, and imagination, and with them, the truly difficult and interesting questions pertaining to the status of our various fictions. To lend credence to a realist view of science makes such a starting point for the humanities wholly untenable. No absolute foundation for sociohistorical research should be sought in an idealist semantics and aesthetics, for the objects of this research are clearly historically contingent events and artifacts.

Yet the adoption of a realist view of the natural sciences still leaves the anthropological and historical disciplines an immense degree of autonomy. This is the case because such a view does not entail any adherence to the eliminative and reductive forms of materialism, that is, those that insist on the reduction of all mental entities, cognitive acts, and acts of human volition to a lower and more basic level of description—for example, a physical one from the vantage point of which all of "the mental" is pure illusion. A nuanced monism does not support any such claim at all and can insist upon the necessity of recognizing the enabling conditions of thought, action, and meaning without denying the possibility of an emergence of properties and capacities not reducible to those conditions.[34] The burden of proof today rests upon the shoulders of those who dream of the ultimate reduction, the central state materialists. Although we

[33]Here I touch upon an interesting issue concerning the evolution of Ingarden's thought. There can be no doubt that before the Second World War he ascribed to idealist tenets that placed the domain of ideas and values beyond the sphere of human creativity. He wrote in a 1939 essay, "Man and His Reality," that man does not create values and that what matters is that man can "attain to that sphere of being which is comprised of values"; in *Man and Value*, trans. Arthur Szylewicz (Munich: Philosophia, 1983), 30 n.5, 25–31. Yet, in an essay that appeared in 1961, "On Human Value" (21–24), the thesis has shifted and man is said to be the creator of values that never have autonomous existence (23). At the same time, it is clear that Ingarden became increasingly interested in a systems-theoretical biology, for he devotes quite a few pages to an attempt to link its conceptions of organismic autonomy to a theory of moral responsibility; "On Responsibility and Its Ontic Foundations," 53–118, esp. 84–100. A detailed investigation of such matters is beyond the scope of the present study.

[34]A valuable discussion of the concept of "emergence" is that of Maurice Mandelbaum, "A Note on Emergence," in *Philosophy, History, and the Sciences: Selected Critical Essays* (Baltimore: Johns Hopkins University Press, 1984), 64–70.

cannot *prove* the impossibility of such a reduction and therefore cannot assert the absolute ultimacy of emergence, we have a strong warrant to assert that explanations within the psychological, cultural, and social levels of description possess a relative autonomy in relation to their physical and biological bases.

The reader may feel, and quite rightly, that none of the particular arguments about the differences between scientific and humanistic research have been dealt with at all in this chapter. Yet a great deal of ground must be covered before it is possible to demonstrate that the word "research" can be applied to both the sciences and the humanities in any strict sense. I have shown, I believe, that the problem of metalanguage is central to any attempt to discuss the relations between humanistic and natural scientific knowledge claims, and I have argued that, although the radicalism of Husserl is in one sense wholly consistent, his failure to establish a plausible "ground" outside the "natural attitude" is quite telling about what such idealist strategies can be expected to yield. This lesson does not, however, amount to any capitulation to the reductive versions of naturalist thought; all that must be relinguished are forms of idealist dogmatism, not any belief in the relative autonomy of the psychological and social levels of description with their "emergent" realities. As the analysis of Ingarden showed, these levels are betrayed when timeless idealities and ghostly aesthetic entities serve as the ultimate anchors for what may be invariant in the practices and interactions of human agents. This difficult lesson has yet to be learned by many humanists and critics, and thus it remains pertinent to point to the broad constraints a naturalistic perspective weighs upon their interpretive efforts.

In the next chapter, the footsteps of my argument lead toward another set of objections to the viability of a scientific knowledge of culture.

CHAPTER 5

Arguments on the Unity of Science

The question of the unity of science has a long and exceptionally complex history. Unfortunately, in literary critical circles this history is frequently reduced to one particular position that may be taken on the topic. For many literary scholars, the idea of a unification of the sciences represents primarily a positivist threat to the autonomy of their own fields. Consequently it is thought that the entire project of unified science must be condemned along with positivism, which is rightly understood to be a wholly unacceptable and dated conception of the natural sciences. Yet as Karl Popper remarked long ago, such hasty conclusions are only warranted if we allow that "positivism" indeed equals "science," which is in fact a simplistic, ambiguous, and wholly dubious proposition.[1] It should not be taken for granted that the idea of a unified science is equivalent to the imposition of positivist epistemologies on research into psychological, social, and cultural realities. It is true that there have in the past been several different positivistic proposals for the reform of the humanities, but these episodes do not justify the humanist in turning away from the lessons, positive *and* negative, of the natural sciences. The more general point here should not be confused with the particular issue of positivism; what is at stake in general is the idea that some basic patterns of inquiry and explanation could be

[1]For a useful survey of the unification mythology of the Vienna circle, as well as a Popperian critique, see Alain Boyer, "L'Utopie unificatrice et le cercle de Vienne," *Cahiers du C.R.E.A.*, 5 (1985), 69–94.

successful in relation to both natural and human realities. Thus, in what follows, what is meant by a "unification" of the sciences is not the extension of any specific positivist philosophy of science into the humanities, nor the goal of arriving at a single, fully systematic body of knowledge—as in the Laplacean myth of the great reduction. What is meant by "unification," then, is the possibility that some of the most basic principles of natural scientific research, as well as some of the physical and life sciences' specific findings, must play a fundamental role in the constitution of any genuine human science. It is clear that everything depends on which scientific principles and findings are being referred to, and on that score I rely upon my previous discussion in which skeptical, relativist, and reductionist sociological views of science are ruled out and a moderate form of realism recommended (further broad stipulations regarding the patterns of scientific explanation are presented in the next chapter). Thus, although in what follows I argue against some of the criticisms of unified science, it should be remembered that what I want to defend is *not* the generalization of a kind of knowledge based upon the myths of scientism, which were flatly rejected at the end of Chapter 3.

In Chapter 4 I sought to display what I take to be some of the typical difficulties inherent in attempts to argue against the very possibility of a unification of the sciences. The example of Husserl was used to illustrate the dilemma facing any *radical* rejection of this kind of unification. First, there are the difficulties met by any thoroughgoing idealism; this kind of belief counters the naturalist unification of science by advancing a unifying program of its own, one that amounts to the universalization of an idealist interpretation of reality. While idealism tenaciously asserts the absolute autonomy of spirit, of the transcendental subject, and of various other unembodied mental and cultural entities, this viewpoint has the problem of explaining the results of the physical and life sciences as a series of spiritual accomplishments, and it is simply not clear how a hermeneutic insistence on prejudices and horizons of expectation is going to explain even the most elementary findings of molecular biology. Yet it is quite obvious that there is nothing in theoretical biology to support the authority of the ancient and venerable texts on the question of the origins and nature of organic processes. Nor can the idealist fiction of the transcendental subject readily explain the physical, biological, psychological, and social conditions of knowledge. In short, the idealist unification cannot succeed. On the other side of our dilemma are the difficulties confronted by the various dualist positions. Foremost among these difficulties is the problem of claiming a significant measure of autonomy for a discipline or program of research that must also acknowledge the validity of natural science in "its own spheres." We are again and again led to ask whether the discourse that assumes the function of assigning limits and circumscribing spheres can ground these claims without relying upon the basic principles and findings of scientific inquiry. If such a discourse does rely upon these principles and findings, the so-called methodological difference thereby tends to vanish or become a matter of degree; it also seems that, if the metalevel discourse does not rely upon science's basic principles and findings, it is in danger of tipping over into some form of idealism—or into one of the varieties of skepticism and relativism criticized earlier in this book.

My goal in the present discussion is to convey my sense of what can and cannot be expected from this type of argumentation, it being my impression that in literary and humanistic circles there are many illusions on this score. Although it is true that a significant amount of literary research is consciously or unconsciously patterned after what is taken to be the model of scientific inquiry, a lot of work is also done in accordance with the idea that humanistic research is in no way guided or constrained by scientific methods or notions. Most literary scholars are somewhat familiar with past attempts to create a literary science, such as those of Hippolyte Taine, I. A. Richards, and the semioticians; many scholars are also well aware that the scientific status of these doctrines is a dubious matter. Yet the failures of such projects do not in themselves provide justification for a global rejection of the goal of unification. In other words, what must be discussed is not the factual issue of whether there presently exists a well-established body of scientific explanations concerning literature or any other social, psychological, or cultural realities; what is at stake here is the *possibility* of arriving at such knowledge; in a second moment, one may also raise the question of the *desirability* of such an achievement. What I surely do *not* attempt in this chapter is a set of detailed methodological and factual guidelines for the project of a human science, chosen from among the welter of competing research programs; rather, my concern is to reveal the problems inherent in some of the prominent objections to the unification project as such, as well as the crippling weaknesses of some of the major alternative models for research in the humanities. That there is presently no unified science of man does not prove that absolutely no progress has been made in human history toward a more realistic knowledge of social, cultural, and psychological realities. Even less does the very real confusion and disagreement within the anthropological disciplines prove that much more progress cannot be made in the future. The positions I criticize in this chapter, however, foreclose on this possibility, and do so as a matter of principle and in an a priori fashion; the views I criticize prefer the pseudocertitude that declares the human sciences to be impossible, to the risks and uncertainties the scientific project necessarily entails.

I do not here enumerate the various thinkers who have argued, in one way or another, for a dualist or separatist position on the issue of the unity of science. A number of typical positions were evoked in the introduction to the third part of this book, and I think it fair to assume that many of my readers are familiar with the work of writers who contend that such domains as language, human history, the psyche, symbols, action, values, and religious experience are simply not susceptible to scientific explanation. The reasons given for this kind of conclusion vary widely, of course, but there are some obvious, recurrent themes, such as that the phenomena in question are irreducibly complex and lack the lawlike and determinate nature that is the anchor for all genuinely scientific explanations. Only a superficial view of the issues can be achieved if we remain at this level of interrogation, and thus I sketch a broader context of analysis before taking up specific examples. The first step to be taken is to guarantee that the analysis avoids the two polarizations that characterize the literature. On the one hand, it is crucial to reject all positivist simplications whereby the contexts of justification and discovery are cleanly sundered. According to this positivist myth, the discussion of science need not, for example, refer to any of the social conditions that make scientific explanations possible. The question of science then becomes a purely logical matter, which means that positivism ultimately shares some key assumptions with the idealisms that it pretends to reject, for its "explanations" and "knowledge" exist in some purely abstract and atemporal space. On the other hand, however, is the opposite danger of a sociological reductionism whereby all epistemological statements about the status of knowledge claims and explanations are replaced by descriptions of social interests and institutions—or even worse, by a far-fetched metaphysics of "forces" and "diagrams" held to constitute all of the so-called discursive formations that determine knowledge.[2]

[2]I refer here to the work of Michel Foucault. Bluntly put, I do not consider Foucault to provide a good model for the literary (or other) disciplines, and this at both a substantive (thematic) and methodological level. Foucault's sweeping historical theses are oriented and supported by an unbridled metaphysical speculation whereby

In order to avoid these two extremes, we need to reformulate in sociohistorical terms our question concerning the possibility of a human science, while also maintaining the epistemological dimension. With this goal in mind, I set forward the following schematic assumptions in order to establish a general context for the discussion

---

they are linked to an inordinate proliferation of mythical entities, such as the "forces," "diagrams," "regimes," and "strata" that are posited as the unseen conditions behind the realities of human history. Foucault is committed to an extreme form of framework relativism, and one can only agree with Gilles Deleuze when he notes that, for Foucault, what matters is not the work of the historian but the search for "conditions"; *Foucault* (Paris: Minuit, 1986), 124. Foremost among the hidden conditions are the mysterious "forces" that apparently generate the various formations under which experience becomes possible. Thus both "Sight and Speech are always already entirely caught up in the power relations which they presuppose and actualize" (*Foucault*, 89). The commitment to this fully historicized ontology leads Foucault to state the most outlandish views on the status of natural scientific knowledge. One may usefully turn to such a work as *La Naissance de la clinique: une archéologie du regard médical* (Paris: Presses Universitaires Françaises, 1963) for examples. Here Foucault purports to describe an epistemic shift or displacement by means of which the "concrete a priori" of something called "positive medical science" came into being. Foucault denies explicitly that this transformation had anything to do with an increase in objectivity, or with anything like a methodological or epistemological progress. Indeed, he speaks of the emergence of modern medicine as a displacement or spatial redistribution by means of which man was constituted as the object of a positive knowledge: "The dark but solid plot of our experience was constituted towards the end of the 18th century" (203). In other words, modern medical science owes its existence and specificity to an "a priori" framework which somehow replaced an anterior framework, and this modern framework has no real epistemic superiority over the formations which it supplanted: the modern is "no less metaphorical," Foucault writes (vii). Foucault thus appears to feel warranted to "analyze" this modern formation in a series of exaggerated metaphors. In the place of descriptions of specific discoveries, the archeology presents a series of broad philosophical assertions about the meaning of the transformation, a meaning couched in an uneasy combination of metaphysical and sociopolitical terms. A telling example is Foucault's treatment of the practice of dissection and the corresponding advances in anatomical research. We learn that, as a result of its investment in the practice of dissection, modern medicine is inscribed within the "discursive space" of the cadaver, which is said to imply an intimate link between this science and the death of its object, a death that is then metaphorically related to the concept of the "finitude" of mankind (200). At the same time, an overarching theme in the book is that the "medical perception" that is being organized toward the end of the eighteenth century emerges alongside a new regime of domination, and the text proclaims this theme in many lyrical passages about the "majestic violence of light" and the domineering and sovereign nature of the "glance" or "look" of science. A cherubic vision of these matters would, of course, be erroneous, but Foucault's descriptions of the awesome regimes, forces, and conditions subjecting other forces to the atrocity of modern medicine strike me as being highly farfetched. It is certain that his contentions are not supported by precise argumentation, detailed documentation, or dialogues with alternative studies of the subject, and it is an error for literary scholars to cite this and similar works as authoritative texts on the matter.

Foucault's work is, of course, subject to divergent readings, and it is possible that his highly creative and provocative writings have served as a spur to some valuable literary investigations. Yet I consider the reading elaborated by Gilles Deleuze to be far

of some particular dualist arguments and positions. Let us assume, then, a view of scientific explanations that is pragmatic and sociological; in other words, we adopt the hypothesis whereby explanations are *not* ideal entities existing in a purely abstract realm; rather, explanations are the products (involving cognitive processes, actions, utterances, physical operations, etc.) of human agents and thus occur in specific historical and social contexts. An explanation involves specific kinds of interactions between particular agents under certain social conditions, relying upon various cultural artifacts, norms, conventions, and so on. Moreover, a successful explanation requires the satisfaction of certain conditions on the side of the reality being explained—a world without regularities, for example, would be inimical to science. In order to simplify the exposition of the argument, one may refer to the myriad conditions behind a given explanation as the "conditions of explanation" (which would thus include the varied psychological, social, and cultural realities without which this explanation would not have been produced). We also assume that the explanations under consideration here satisfy some minimal requirements imposed by a realist conception of science, namely, that these explanations (at least partially) refer to a state of affairs existing independently of the scientists' explanatory theories and statements, and also that a given state of affairs is explained by being linked to the anterior nomic conditions that are the reasons for its being the case (more details on this topic are advanced in the next chapter). Next I adopt my previously discussed thesis that the natural sciences have in fact produced some genuine explanations in this sense. We can thus refer to the conditions that, having been satisfied,

---

more faithful to Foucault than any selective and reasonable appropriations could be. Deleuze is right, then, to emphasize the affinity with Alfred Jarry's pataphysics, and he also correctly highlights the extent to which the avant-garde fiction of Raymond Roussel was a kind of epistemological model or pattern for Foucault—a point also made in a review by E. S. Schaffer, "The Archeology of Michel Foucault," *Studies in the History and Philosophy of Science*, 7 (1976), 269–75. Deleuze cites a phrase in which Foucault claims that he has never written anything other than fiction, which seems somewhat apt, but I cannot agree with Deleuze's ensuing remark according to which no fiction has ever produced so much truth and reality (*Foucault*, 128). Deleuze is closer to the mark when he highlights the centrality of the following passage, which is said to define the two poles between which Foucault incessantly oscillates: "From among so many things having no status, and from among so many fantastic civil statuses, [Leiris] slowly collects his own identity, as if in the folds of words there slept, along with chimeras that never really die, absolute memory. Roussel brushes aside these same folds with a concentrated gesture, in order to find in their place an unbreathable vacuum, a rigorous absence of being, which he can dispose of in complete sovereignty, to fabricate figures having neither kin nor kind"; Foucault, *Raymond Roussel* (Paris: Gallimard, 1963), 28–29; cited by Deleuze, *Foucault*, 106 n. 12.

made possible these genuine explanations of aspects of natural reality.

Given these terms, it is possible to formulate the issues related to the unification of science on a nonpositivist basis. The most general question concerns whether there is any reason why the *kinds* of conditions which make the production of natural-scientific explanations possible could not be satisfied in regard to human, cultural, and social realities. To focus in this manner on the conditions of possibility of knowledge is an improvement because it allows us to identify the different topics that typically are emphasized in the formulation of a dualist position. Some dualists focus on the purported inapplicability of the basic model of explanation to anthropological realities on the grounds that the nature of these realities is somehow inimical to explanation—this being a claim about the ontology of "the human." Other dualists stress the historical and social background that is supposed to alter the basic status of the knowledge we may have of human affairs. There are also those who take up what I call the "reflexivity" argument, which suggests that a human science would require an impossible feat of self-reference in order to succeed. In each case, then, reference is made to the impossibility of satisfying the conditions under which successful (that is valid) explanations of natural realities have been produced.

The core of my strategy, which could be applied to any number of different dualist arguments, is to analyze these positions with the goal of isolating their fundamental conceptual or factual bases. The question to be posed, then, is whether these viewpoints do not finally rest, in a manner that can be shown to be crucial to their validity, on one or more of the theories of knowledge I have previously discussed, namely, an incoherent form of skepticism, a kind of framework relativism (e.g., radical sociological reductionism), or idealism (e.g., belief in the ontological autonomy of mental substances). Insofar as I consider that these positions have been shown to represent wholly inadequate and invalid theories of knowledge, it follows that they hardly provide any true support for specific contentions about the possibility or impossibility of a human science. In what follows, I indeed set out to demonstrate that some of the most influential and prominent dualist arguments can in fact be undermined by means of the strategy just outlined. The question then remains whether there are any dualist arguments that are not, at base, applied versions of skepticism, framework relativism, or idealism (semi-or integral). I cannot of course prove that there are *none* or never could be any; my claim is that the possibility is unlikely insofar as dualist arguments that skirt these pitfalls typically do so by moving closer and closer to a realist conception of science. Should this be the case, the dualism of methods and objects is replaced by a matter of degree, amounting finally to a differentiation of research and findings within a broad, overall scientific project, the unity of which is anything but a constricting dogma.

I begin with a discussion of some of the dualist arguments typically associated with Ludwig Wittgenstein. This is an appropriate starting point in the present context because Wittgenstein's writings have frequently been thought to lead the way toward a precious reorientation of our views on the problem of the nature and status of our knowledge of such specifically human realities as language, meaning, value, works of art, and ethical and religious beliefs. In the typical version presented of the later Wittgenstein's message, the main contribution that his analyses make to the humanistic domains is to show the necessity of some kind of dichotomy whereby the "open" concepts and "participatory" modes of understanding, which are held to be necessary in the humanities, are contrasted to the "sharp" definitions and objectifying theoretical explanations provided within the natural sciences. I begin with a brief recapitulation of this typical Wittgensteinian view before going on to show that the situation is far more complicated, both in Wittgenstein's texts and in reality. A second reason why it is appropriate to begin with a commentary concerning matters Wittgensteinian is that many of these ideas have had a particular appeal within literary circles, and it could be useful to point to some of the many real difficulties inherent in a philosophy that is sometimes pointed to as the source of any number of major solutions. I think that anyone who has taken the trouble to look closely at the problems within Wittgenstein's philosophy will conclude that to speak of a badly needed Wittgensteinian reorientation of literary criticism is at once totally vague and highly misleading; in this regard, literary studies are at the present moment lagging behind a previous development within the philosophy of the social sciences, psychology, and language, where sweeping promises of the past have been replaced by a recognition of the real complexity of the issues that Wittgenstein usefully raised but did not settle.

Wittgenstein and the Craving for Generality

I begin this discussion by taking up ideas that hover at the boundary between what Wittgenstein actually wrote (and may at some point have believed) and what has become a kind of standard account that circulates in the literature.[3] Arguments arising from either side of that boundary are of interest in our context. The first and most basic point urged by Wittgenstein's followers is that human languages and cultures involve a large constellation of terms that do not represent definite extralinguistic realities. These expressions, which typically include such notions as meaning, understanding, ethics, and aesthetics, are thought to be fuzzy and to have no essential sense or reference. Such concepts are, on the contrary, only "family resemblance" terms. In practice, we manage to have workable, operative definitions of them. In other words, in a given context—and as a result of a practical, participatory form of knowledge—we can usually decide whether we want to include a given instance beneath one of these terms (whether, to use Wittgenstein's own classic example, we want to consider a given activity a game or not). Thus we have what Leibniz called "nominal" definitions, but we do not have "real" ones, for we cannot give an exhaustive statement of the necessary and sufficient properties of such entities. This fact is thought to entail the idea that we should be content to give up looking for reductive, essential definitions and explanations in regard to such terms; rather, knowledge of them is a matter of their proper use in the given context; at most, one could undertake careful descriptions of the diversity of these uses—it being understood that in the absence of any overarching laws or regularities governing them, the description of such uses will necessarily be open-ended, partial, and provisional.

The effect of this insight is allegedly great, for it would permit us to put to rest any number of questions and problems that have been the object of numerous, pointless debates. Instead of seeking for more and more rigorous and exact explanations of cultural realities—always on the model of investigations in chemistry and anatomy—we should realize that such interrogations break with actual linguistic practices or "ordinary language" usage. Thus Wittgenstein accuses the Platonic and metaphysical philosophical traditions, as well as the scientific spirit, of importing the wrong kinds of questions into the cultural sphere, beginning with the Socratic question par excellence, "What is the Good?" In this sense Wittgenstein would be exemplary most of all for the antitheoretical strains within the humanities, for the central thrust of his work would be to deny the

[3]A useful bibliographical tool is Guido Frongia's *Guida alla letteratura su Wittgenstein* (Urbino: Argalia, 1980).

pertinence and validity of any radical inquiry into foundations, guiding principles, and essences.

It may be interesting to note in passing that, although Wittgensteinians are fond of castigating Platonism, one of their most basic tenets is touched upon by Socrates in a passage of the *Phaedrus* that strongly anticipates the notion of family resemblance concepts. Socrates claims that there are two basic kinds of words. When we use one kind, everyone understands the same thing or thinks of the same sort of object. Such is the case with words like "iron" and "silver," he adds. Yet in other cases we are at variance, individually and as a group, over the meaning of a word. Given that rhetorical devices have the greatest power in those cases where a term's meaning is varied, Socrates proposes to Phaedrus that "he who intends to develop an art of rhetoric must first make a methodical division and gain a clear impression of the distinctive character of each kind—that about which the many necessarily diverge, and that in which they do not."[4] The Wittgensteinian, then, would be someone who has already made this methodical division, and who can thereby tell us which of the concepts related to mankind belong in the class of open concepts, and which other terms do not.

When one tries to grasp what is being said about humanistic concepts within a Wittgensteinian framework, it is helpful to get a better sense of their characterization by asking about the kinds of concepts to which they are contrasted. If the key insight here for the literary critic is that certain crucial critical concepts or notions are really open or vague family resemblance terms, then we want to know several things: (1) are there also supposed to exist closed and exact non–family resemblance terms—and what are they like? (2) if any of the latter do exist, are there some that have to do with cultural or social realities? On both of these points there is a great deal of interpretive dissent, and it would seem rash to pretend to state the "Wittgensteinian perspective" in a direct and univocal manner. It is simply not clear to what extent Wittgenstein really believed that sharp or exact concepts could be obtained with any ultimacy in the natural sciences. A passage in the *Blue Book* flatly states that in the case of a treatise on "pomology" there exists a natural kind against which to measure the adequacy of a definition, whereas in relation to the notion of "game" there does not. Yet some interesting remarks in later works go far toward challenging what may only be the traces of the sharp, Vienna circle dichotomy that was expressed in the *Trac-*

[4]Plato, *Phaedrus* (London: Heinemann, 1914), 263b.

tatus and in the positivistic lecture on ethics.[5] Fortunately, we do not need to provide here an account of Wittgenstein's views on the philosophy of the natural sciences, an issue on which the commentaries are especially fragmentary and inconclusive. The essential point is that whatever Wittgenstein's views on the specific nature of scientific knowledge were, there can be no doubt that he frequently insisted that philosophy—and by extension, other discourses dealing with culture and language—must not imitate the methods of the natural sciences.[6] His proposals concerning the notion of family resemblance terms make most sense within such a context, for it is a matter of suggesting that an illegimate extension of the scientific method into the cultural sphere results in fruitless debates as well as metaphysical illusions. A text that is especially clear and direct on this point is the following, again taken from the *Blue Book*:

> Our craving for generality has another main source, our preoccupation with the method of science. I mean the method of reducing the explanation of natural phenomena to the smallest possible number of primitive natural laws; and, in mathematics, of unifying the treatment of different topics by using a generalization. Philosophers constantly see the method of science before their eyes, and are irresistibly tempted to ask and answer questions in the way science does. This tendency is the real source of metaphysics, and leads the philosopher into complete darkness. I want to say here that it can never be our job to reduce anything to anything, or to explain anything. Philosophy really *is* "purely descriptive."[7]

In what follows I contend that these kinds of methodological strictures, which echo in many ways the kernel positions of the neo-Kantian dualisms, are directly linked to the problem of metalanguage discussed in regard to Husserl in the previous chapter. What kind of knowledge is it, we should want to know, that can accurately trace the limit to the viability of the "method of science"—and what is its relation to the "purely descriptive" forms of knowledge that are being prescribed in relation to such phenomena as games and cultural practices? To ask such a question is not a matter of imposing as if

[5]Ludwig Wittgenstein, *The Blue and Brown Books* (New York: Harper, 1960), 19. "A Lecture on Ethics," *Philosophical Review*, 74 (1965), 3–12.

[6]See P. M. S. Hacker, *Insight and Ilusion* (Oxford: Clarendon, 1972); B. F. McGuinness, "Philosophy of Science in the *Tractatus*," in *Wittgenstein et le problème d'une philosophie de la science*, ed. Anon (Paris: C.N.R.S., 1970), 9–20; Derek L. Phillips, *Wittgenstein and Scientific Knowledge* (London: Macmillan, 1977).

[7]*Blue Book*, 18.

from above some kind of abstract epistemological and theoretical problem—as a result of some habitual attachment to that kind of metaphysics. Rather, such problems are encountered as soon as we start trying to *apply* or follow these kinds of methodological strictures. Should a literary critic, for example, decide that there could be something valuable in the notion of family resemblance concepts—as indeed she or he should—how is it to be determined *which* critical notions should be regarded in this manner? Is there a method, or even a set of criteria, which could be used to approach this issue? Such problems, although of real theoretical pertinence, are not purely abstract, for they emerge within the most particular and descriptive (or "idiographic") work within such a discipline as literary criticism, where one must indeed deal with the question of the "openness" or "closedness" of terms—generic terms, for example. And what seems puzzling about the Wittgensteinian line is the way it seems to presuppose that one possesses an adequate answer to problems it in fact does not answer at all. Let us see how this works in relation to the "open" concept example, which is, of course, but one aspect of the Wittgensteinian inspiration.[8]

As we saw above, the *Phaedrus* refers to a methodic division between sharp and fuzzy terms, or at least between homogeneous and disparate usages. Wittgenstein does not say that a truly systematic division of this sort is possible, no doubt because that would be like granting that an explanatory science of rhetoric were possible. Yet is not the Platonic text more deep in this regard? For even if I agree with Wittgenstein that it would be more prudent and careful to proceed example by example, how am I to know, given the "particular case" of the word "game," that there is no essential definition of this term and that I should treat it as an open concept? This is in fact quite an imposing task, because what would have to be shown is that it is necessarily the case that no logically closed definition of a given term can ever be provided. But how could such a demonstration be achieved? The Wittgensteinian can, of course, point to the long and detailed record of past disagreements and failed definitions. Thus Morris Weitz, in order to make this kind of point in the realm of literary criticism, goes to great lengths to show that the long history of previous definitions and interpretations of such entities as *Hamlet*

[8]Here I am in full agreement with Charles Altieri, who says that to emphasize the "family resemblance" notion is to miss the real point of the Wittgensteinian inspiration; see his *Act and Quality: A Theory of Literary Meaning and Humanistic Understanding* (Brighton, Sussex: Harvester, 1981), 186–87 n. 3.

and "tragedy" are wildly at variance.[9] Yet although such a discussion may be highly persuasive for some readers, it hardly amounts to a conclusive demonstration. It shows us that people have been inconsistent as a group in wielding certain terms, but this is still no proof that they *must* disagree in the future. Even less does it prove that all individual definitions of tragedy are bound to be wrong. When was it established that knowledge required unanimity? Moreover, the Wittgensteinian's apparent insistence on consensus or agreement, and on the need for theory to adhere to the ordinary usage, is not really the key point. Have not people often agreed that it *did* make sense to try and define these fuzzy concepts? Have they not agreed that these terms actually referred to some definite reality, going on to waste a lot of time believing in false essences? The Wittgensteinian does not hesitate to contradict them in this regard. It turns out, then, that the Wittgensteinian is not only making a statement about how people have used words but is making a much more strong and direct thesis, one having to do with the referents of different kinds of terms. And this thesis is, quite simply, that these referents do not exist, that they are fictive or illusory. When the critic or philosopher who has been contaminated by the craving for generality and the method of science tries to find a rigorous definition of some cultural entity, this is a metaphysical error. Why? The only consequential answer to this question can be that no such thing really exists. Here we see that the supposed methodological stricture appears to have a much deeper basis in certain ontological theses. Thus what is really in question is the drawing up of a list of terms that should no longer be thought to refer to essences or to similarities between particular substances; it is a matter of deciding that there is no reality outside of language, no "species" of apples, which could serve as the basis for a rigorous or closed definition of these terms. This is so because there are no such common and invariant properties "out there" waiting to be discovered in the case of such family resemblance concepts as games; particular games may very well exist as the specific activities of human beings, but no common nature exists to serve as the essential reference of the general term "game." In other words, the grammar of the word "game" cannot be defined univocally attending to the word's *reference*; rather, the grammar or use of such a word is "essen-

[9]Morris Weitz, *Hamlet and the Philosophy of Literary Criticism* (Chicago: University of Chicago Press, 1964). This work is rightly criticized by John Ellis as failing to grasp many of the nuances of Wittgenstein's views, in *The Theory of Literary Criticism* (Berkeley: University of California Press, 1973). The same is true of Ellis's book.

tially" a matter of the *disparate* practices and activities that one finds within the so-called forms of life or, speaking more loosely, language games. But how do we know that these practices are really so very disparate? How do we know, for example, that the diversity of human games is not adequately embraced by the four categories proposed by Roger Caillois in *Les Jeux et les hommes*? In other words, the epistemological question returns to haunt the ontological theses of a proposal that is supposed to be of methodological value.

Here an issue that I touched upon earlier reappears. Supposing that we accept this much of the Wittgensteinian line, we still need to know where that line is to be drawn. Are all literary and humanistic terms to be viewed in this way? Surely it cannot be a matter of proposing to us a methodology in which all general terms are banished from descriptions of cultural practices because these practices are deemed to be really a matter of so many particular entities (and events?) bearing no real similarities to each other. If that is what the Wittgensteinian believes, how could he or she speak to us coherently about such general words or terms as "meaning," "value," or "literary criticism"? Moreover, the Wittgensteinian is clearly committed to the consistency and validity of various general, referential terms having to do with human beings and their use of language. Wittgenstein himself relies upon certain general terms of this sort, such as *Gebrauch*, *Institution*, *Situation*, *Darstellungsweise*, and the notorious *Lebensform*. Thus the methodical division being proposed cannot be that only general propositions referring to the common properties of physical particulars can have truth values and make sense, while all general terms or propositions referring to common traits of particular human entities are in fact nonreferential and imaginary. Rather, the methodological and ontological division must be traced somewhere *within* the sphere of human signification and cultural realities. That step seems quite correct, for it does seem to be plausible to assume that not all of the general substantives used in literary discussions, for example, really refer to transcultural universals. At the same time, it seems like a good rough start for social and cultural science to begin with the assumption that there have never been human beings who have managed to live together without engaging in practices or behavior revealing regularities or recurrent facets; one may debate at great length over the status of such patterns and the best general terms to be used to designate them, such as "role," "position," "norm," "type," "structure," "function," "institution," or "rule." In spite of the actual diversity of particular historical realities, and in spite of the difficulties surrounding the various general terms that one may use to map them, it should be clear that one should not try to avoid these general concepts as a result of some ontological phobia.

Does the notion of family resemblance take us any further than this? The valid intuition behind this term is that as long as one fails to find a rigorous and comprehensive definition of a concept, as long as the set of particulars it picks out cannot be decided in terms of a single law, pattern, or set of criteria, one surely has no reason to conclude in advance that this concept designates an essence that definitely exists and is waiting to be found. But does this valid intuition imply that in such a situation one may draw the opposite conclusion, namely, that no regularities or common properties corresponding to the concept exist? I think not, yet in fact literary critics influenced by the Wittgensteinian strictures have often leapt to such conclusions in a very general way; we are told, for example, that it is simply not part of the grammar of critical discourse to refer to anything.[10] It would be far more prudent to consider that there is no prophetic method for predicting the future of the distinction between sharp and vague terms, between terms that refer to invariant kinds and terms that do not; it seems particularly disastrous, within the humanities, to make any such preemptive assumptions. As a heuristic principle, it may be more useful to keep in mind the possibility that many general humanistic concepts may not ultimately refer to any universal or invariant properties of human thought and action; as a result the human scientist must be very prudent about which general concepts guide the inquiry into history and culture. In this regard, David Bloor's way of employing the family resemblance notion is perceptive, for he reads it as a heuristic injunction against a dogmatic belief in the existence of ultimate essences. He claims that the "injunction to banish family-resemblance concepts is really the injunction to stop doing science."[11] But of course this insight is two-sided, for we must also note that the injunction to stop trying to define, classify, categorize, subsume, and explain is also an injunction against science, which, if it requires a critical and nondogmatic attitude, is also

[10]An argument along these lines is that of Suresh Raval, *Metacriticism* (Athens: University of Georgia Press, 1981), a work explicitly situated within a Wittgensteinian lineage. The author asserts that the "kind of question that literary criticism raises is radically different from that raised by scientific inquiry" (159). A key reason for this seems to be that literary criticism is not a problem-solving activity in the sense of an inquiry concerning issues in relation to which there are evidence and answers (160, 181).

[11]David Bloor, *Wittgenstein: A Social Theory of Knowledge* (London: Macmillan, 1983), 37.

suffocated by the kind of skeptical a priori that some people derive from Wittgenstein. In any case, the epistemological insight that Bloor finds in the family resemblance concept is in no way particular to the humanities, for natural science has always had to deal with the same problem.[12] Thus there is no basis here for defending the overdrawn neo-Kantian dichotomy between idiography and nomothesis, that is, between the particularist and descriptive approach of the humanities and the reductive and explanatory (lawlike) method of the sciences.[13] There is no basis, because the viewpoint being criticized depends upon a wildly skeptical conclusion about the reference of general humanistic concepts; this skeptical conclusion in turn rests upon a sweeping and dogmatic ontological thesis concerning the absolute particularity of human and cultural entities. Moreover, the position in question is incoherent in that it necessarily employs the kind of general terms it would banish from the discourse of the humanities.

As I mentioned from the outset, it is sometimes important to distinguish between a kind of typical Wittgensteinian viewpoint and the various nuances and perspectives of Wittgenstein's highly complex texts. On the particular issue of family resemblance, this kind of scruple may be appropriate, as literary critics have clearly made their own use of the notion. Some of Wittgenstein's most careful and untiring commentators have pointed out that the discussion of family resemblance in the *Philosophical Investigations*, far from being a general, methodological proposal for a way of doing philosophical analysis, is best understood as being purely negative in its thrust—even if some of Wittgenstein's own remarks *do* defend theses that are more positive, and that are, in fact, indefensible. According to G. P. Baker and P. M. S. Hacker, Wittgenstein should not be taken to mean (and should not have said as he in fact did at *Philosophical Investigations* § 65 and elsewhere) that games have no common properties; rather, his purely negative thesis would be that we need not look for the positive properties possessed by all games: "The essential points are that we know of no properties common to all games; that we do not explain "game" by enumerating *Merkmale* of games; and that even if we were to discover a property common to all games, it would

[12]Bloor does not read Wittgenstein as a dualist and contends that the history of the concepts of the natural sciences reveals an oscillation between vagueness, sharpness, and the disorder and anomalies that will always arise in order to disturb any organizing grid; Bloor, *Wittgenstein*, 33–37.

[13]Anyone in search of arguments against such a dichotomy could usefully begin with Kant himself in the *Critique of Pure Reason*, B 680–92.

not reveal part of our concept of a game because it would not belong to our (present) practice of explaining 'game'."[14]

These statements take us to the heart of the difficulties, confusions, and ambivalences that surround the very idea of finding methodological guidance and general theoretical implications in Wittgenstein's remarks. Again and again, we find that Wittgenstein's voice is implicitly situated at a methodological or metalevel of analysis, a level at which the possession of a systematic, general, and exhaustive overview or survey of the issue in question is implied. This is the level that Wittgenstein adopts overtly and clearly when he rashly states that we can "see" that there are no common features in all games, and that it would thus be wrong to go on looking for an artificial, scientific definition whereby the word could be used as an exact term. Again and again, we find in Wittgenstein any number of disclaimers about the possibility and desirability of situating oneself at such a level of generality, which is, after all, to succumb to the same craving that was earlier denounced. Thus the thesis of family resemblance is said to be purely negative and only a critical objection to someone else's postulations (Wittgenstein as the gadfly's gadfly). Yet this interpretation is incorrect, for we can clearly see in Baker and Hacker's remarks that the negative statements are pronounced in regard to a reality that it is Wittgenstein's goal to describe and to defend. Insofar as Wittgenstein's work has any real methodological strictures or insights, they have to do with what he says about the general kinds of intellectual attitudes that he thought were appropriate and inappropriate to that reality. This reality, which it is the goal of Wittgenstein's purely negative thesis to defend, has many names in Wittgenstein's work. A good first indication of its nature is provided by two of Baker and Hacker's expressions in the above citation: "our concept," and "our (present) practice."

Forms of Life, Forms of Knowledge

Here we encounter the well-known theme of Wittgenstein's defense of the "grammar" of ordinary language—its actual usage in everyday affairs—from the wrongheaded incursions of scientifically minded philosophers who, beginning with Frege and Russell, sought,

[14]G. P. Baker and P. M. S. Hacker, *Wittgenstein: Meaning and Understanding* (Oxford: Blackwell, 1983), 192–93.

in a wholly artificial manner, to carve an object called "language" out of the complex tissue of human practices. Once separated in this manner, the artificial object called "language" could then be purified and reduced to a lawlike calculus, a bundle of tokens and a set of rules for making well-formed strings out of them. The theme of Wittgenstein's defense of actual "grammar" against this kind of formalized theory is indeed a central and valuable facet of his work, yet to pursue it fully is to realize that his rejection of the formalist and scientific theory of language is not solely negative insofar as it by the same stroke amounts to a defense of something else. Many different terms may be used to refer to this "something else," and to use the term "our present practice" is wholly appropriate. There is, however, some consensus in the literature—and rightly so—that the ambiguous concept of *Lebensform* is an "ultimate" for Wittgenstein: it is the forms of life, and not language games or institutions, that are the constitutive features and frameworks of what other theoretical discourses single out as linguistic or semiotic practices—at the cost of ruinous abstractions. Wittgenstein's position, then, does not only amount to the negative gesture of castigating philosophy's illusory departures from the basic forms of life—its mistakes about "our concepts" and "our (present) practices"; a truly Wittgensteinian methodology for literary studies or for the humanities in general must also adopt the way Wittgenstein thinks these *Lebensformen* can effectively be known (and *are* in fact known all the time). Wittgenstein may very well reject the scientific model of knowledge, but he does not adopt a generally skeptical attitude. Although he wrote a series of remarks criticizing G. E. Moore's refutation of academic skepticism, it was not in order to defend the latter.[15] On the contrary, Wittgenstein thought that Moore's manner of refuting skepticism gave too much away to its opponent, and thus such a defense failed to bring forth the irrelevance of skeptical doubts. That Wittgenstein was heavily invested in defending the certainty of some types of knowledge is implicit in Baker and Hacker's expressions, for they

[15]Skeptical doubts are discussed by Wittgenstein in *On Certainty*, one of the leitmotifs of which is the idea that doubt too has its presuppositions; *On Certainty*, ed. G. E. M. Anscombe and G. H. von Wright (New York: Harper & Row, 1972). Citations to *On Certainty* hereafter refer to the number of the remark (not page) in the above edition. Three excellent commentaries on this work are Thomas Morawetz, *Wittgenstein and Knowledge: The Importance of "On Certainty"* (Amherst: University of Massachusetts Press, 1978); W. D. Hudson, "Language-Games and Presuppositions," *Philosophy*, 53 (1978), 94–99; and G. H. von Wright, "Wittgenstein on Certainty," in *Problems in the Theory of Knowledge*, ed. G. H. von Wright (The Hague: Nijhoff, 1972), 47–60.

oppose a common, perfectly adequate practice of explanation to one held to be at once inappropriate and unsuccessful. It is this common practice of understanding and knowing that is being defended whenever the Wittgensteinian formulates arguments against the possibility of a natural-scientific investigation within the human sphere.

Wittgenstein, then, does not oppose the project of a unification of science on purely skeptical grounds, if by this is meant a belief that we can have no certainty or knowledge in regard to human and cultural realities. On the contrary, his position on this issue amounts to two major claims. The first, which we have just reviewed under the usual rubric of "family resemblance" terms, involves arguments denying the possibility of successfully arriving at formalized and lawlike explanations of anthropological realities, beginning with language. These arguments manifest a local skepticism concerning the possibility of nomic generalizations in the humanities, and as I have contended above, this skepticism in fact rests upon some very strong (and ultimately incoherent) ontological theses. The second major claim is an apparently more positive contention concerning the kind of knowledge we have of human affairs in the absence of the "puzzles" brought by philosophico-scientific inquiry, on the one hand, and academic doubt on the other. Bluntly put, Wittgenstein tends to claim that the legitimate questions that may arise in regard to language, meaning, culture, and so on can be dealt with adequately by clarifying the nature of "our present practices" (forms of life). Furthermore, this clarification can be achieved through a careful *description* of "our usages" and does not require a full-fledged "explanatory" effort.

What makes it difficult to give a more detailed and systematic characterization of this alternative mode of knowing, however, is the fact that Wittgenstein explicitly denies that the kind of knowledge he advocates is in any way systematic, that it can ever be made fully explicit, or that it can be justified by any basic principles. "At some point one has to pass from explanation to mere description," Wittgenstein wrote,[16] and the central question concerning the kind of certainty he proposes in such remarks concerns what it is that these descriptions can be taken to describe. The short answer is, of course, "our forms of life," but the truth of the matter is that this slogan merely crystallizes the difficulties and ambiguities surrounding Wittgenstein's variety of certainty. It is to these difficulties and

[16]*On Certainty*, § 189.

ambiguities that we turn now. It should be noted in advance that one may distinguish between two facets of the problem. Wittgenstein advances certain substantive theses about the *nature* of the forms of life and about the practical role of knowledge (or more accurately, cognition) within these forms. But he also advances claims at another level, which is properly epistemological, and these claims concern what we can and cannot know about these forms, the nature of this knowledge, and its relation to other forms of knowledge. These two levels are clearly interconnected, but it is important to note their differences as well, and I emphasize the extent to which those who want to pull methodological guidelines or models out of Wittgenstein's writings should focus on the issues related to the latter level. I take the two up in turn, without, however, striving to keep them absolutely separate in my exposition.

In a sense, little work remains to be done on the issue of the concept of the *Lebensformen* in Wittgenstein's philosophy. Every appearance of the term in the extant writings has been plotted, and the different contexts of its usage have been carefully weighed in an effort to disambiguate the term's sense. The result is that it has been argued quite strongly that the term is marked by a profound ambivalence and hovers between conventionalist or culturalist connotations ("forms"), on the one hand, and some kind of naturalistic or organicist sense on the other ("of life"). Yet in the absence of a more telling account of Wittgenstein's philosophy of nature, it is not wholly clear what is meant by "life" in his usage of the term, to what extent he accepts for example, the concept's romantic (organicist and vitalistic) background.[17] Commentators who wish to stress the strengths of Wittgenstein's philosophy and who are also aware of the shortcomings of the views of such *Lebensphilosphen* as Oswald Spengler (whom Wittgenstein cites favorably) tend to argue for a conventionalist or institutional way of construing the forms. Such a decision does not maintain the integrity of Wittgenstein's writings, for as we shall see in a moment, it simply cannot be denied that his ambivalent conception at times swings resolutely toward a pseudo-biological understanding of the ultimate framework of our knowing

[17]For a survey of the literature on the term in Wittgenstein, see Nicholas F. Gier, *Wittgenstein and Phenomenology* (Albany, N.Y.: S.U.N.Y. Press, 1981), 17–32. Of particular value are J. F. M. Hunter, "Forms of Life in Wittgenstein's *Philosophical Investigations*," *American Philosophical Quarterly*, 5 (1968), 233–43; and Max Black, "*Lebensformen* and *Sprachspiel* in Wittgenstein's Later Work," in *Wittgenstein and His Impact on Contemporary Thought*, ed. Paul Weingartner and Johannes Czermak (Vienna: Hölder-Pichler-Tempsky, 1978), 325–33.

and acting. Yet there is no reason in principle why literary critics and other humanists could not operate with a neo-Wittgensteinian model in which none of the "biological" elements are maintained, and thus one may usefully ask what such a model would be like. On this score I would argue that Wittgenstein does not really give us much in the way of insights into the nature of social conventions, institutions, norms, and so on, and even less does he say anything about the conditions under which these "frameworks" of human life can and cannot become known to the agents who move and think within them. Nowhere does Wittgenstein discuss the fact that the truth of human practices and institutions was not given from the beginning of human history but has only been glimpsed as the result of a long historical process. The basic sociological apperception whereby individuals look upon their own norms and practices as norms and practices—and not as sacred and sacrosanct ultimate realities handed down from a transcendent source—was not "always already there"; nor is it correct to say that our recent history's unveiling of the lack of a transcendent or sacred foundation for social practices is by any means complete. Many of the *Lebensformen* that shape social existence in the modern world—for example the myriad significations, norms, motivations, and practices related to the international monetary system—are not simply lying before us on display, waiting for a purely descriptive knowledge to delineate them. Perhaps the Wittgensteinian response would be that these practices in fact function without the agents having any systematic lucidity about their ultimate nature, that in a practical sense these forms are "in order," and that it is only an "external," scientific or philosophical curiosity that asks for a more essential or more general explanation of the true nature of the economic system. It could be argued that many of the present forms of life (or better, social institutions) maintain their order precisely because the agents are unaware of their nature and misrecognize them for something else.

As was asserted above, Wittgenstein's own statements on the nature of the forms of life are ambivalent, and at times the emphasis clearly shifts to a quasi-biological manner of understanding the concept, which is not the least bit surprising given the tradition out of which the term arises and Wittgenstein's familiarity with nineteenth-century German philosophy. A first example may be drawn from *On Certainty*: "Now I would like to regard this certainty, not as something akin to hastiness or superficiality, but as a form of life. (That is very badly expressed and probably badly thought as well.) . . . But that means I want to conceive it as something that lies beyond being justified or unjustified; as it were, as something animal (*also gleichsam als etwas animalisches*)."[18] Unless we are prepared to say that Wittgenstein thought that the existence of animals was ordered by conventions and institutions, it is clearly wrong to claim that he consistently held to a conventionalist understanding of the concept of "forms of life." On the contrary, the citation suggests that in his effort to situate his notion of certainty beyond or beneath the philosophical problem of justification, Wittgenstein at least temporarily toyed with a biological solution in which a deep and somehow more basic *natural and organic* level of description provides the anchoring for certainty. In this he would resemble Hume's naturalist response to skepticism as well as more recent trends of naturalized epistemology. Yet we see in a moment that the threads of Wittgenstein's position are highly tangled. The context of my next example of Wittgenstein's biological motifs is a series of remarks on Frazer's *The Golden Bough*.[19] Wittgenstein is sharply critical of what he understands to be Frazer's radical misconception of the nature of mythol-

[18]*On Certainty*, §§ 358–59.

[19]These remarks, which were probably written off and on between December 1930 and as late as 1948, were first published in the original German as the "Bemerkungen über Frazers *The Golden Bough*," *Synthèse*, 17 (1967), 233–53. An extremely unreliable English translation, tendentiously edited by Rush Rhees, first appeared in *The Human World* in 1971. John Beversluis's integral and excellent English translation appeared in C. G. Luckhardt, ed., *Wittgenstein: Sources and Perspectives* (Ithaca: Cornell University Press, 1979), 61–81. All further references to the "Remarks on Frazer's *Golden Bough*" give page numbers to the latter edition, referred to as *Remarks on Frazer*. Supplementary sources for Wittgenstein's ideas on Frazer during the 1930s are available in G. E. Moore, "Wittgenstein's Lectures in 1930–33," in Moore, *Philosophical Papers* (London: Allen & Unwin, 1959), 315–16; and Alice Ambrose, ed., *Wittgenstein's Lectures: Cambridge, 1932–35* (Oxford: Blackwell, 1979), 33–34. For some interpretations focusing specifically on *Remarks on Frazer*, see Richard H. Bell, "Understanding the Fire-Festivals: Wittgenstein and Theories in Religion," *Religious Studies*, 14 (1978), 113–24; Jacques Bouveresse, "L'Animal cérémoniel: Wittgenstein et l'Anthropologie," in his edition of the French translation, *Remarques sur le Rameau d'Or de Frazer* (Montreux: L'Age d'Homme, 1982), 39–124; as well as the extended discussion by Bouveresse in his *Wittgenstein: La Rime et la raison* (Paris: Minuit, 1973), 205–34; Frank Cioffi, "Wittgenstein and the Fire-Festivals," in *Perspectives on the Philosophy of Wittgenstein*, ed. Irving Block (Cambridge, Mass.: M.I.T. Press, 1981), 212–37; Jens Glebe-Møller, "Marx and Wittgenstein on Religion and the Study of Religion," in *Wittgenstein and His Impact*, ed. Paul Weingartner and Johannes Czermak, 525–28; Elisabeth List, "Zum Problem des Verstehens fremder Kulturen: Wittgensteins Bemerkungen zu James G. Frazers *Golden Bough*, in *Wittgenstein and His Impact*, 471–74; Anthony O'Hear, "Wittgenstein's Method of Perspicuous Representation and the Study of Religion," in *Wittgenstein and His Impact*, 521–24; Rush Rhees, "Wittgenstein on Language and Ritual," in *Wittgenstein and His Times*, ed. Brian McGuinness (Chicago: University of Chicago Press, 1982), 69–107; N. Rudich and M. Stassen, "Wittgenstein's Implied Anthropology," *History and Theory*, 10 (1971), 84–89.

ogy and ritual. Whereas Frazer tries to comprehend these phenomena in a direct comparison to science and technology, Wittgenstein responds that this is a confusion of frameworks of analysis. In fact, the "primitives" do not mistake their ceremonial practices and myths for literal truths about the works of the natural world; rather, these practices and myths have a wholly different basis and meaning. Thus Wittgenstein writes:

> That is, one could begin a book on anthropology by saying: When one examines the life and behavior of mankind throughout the world, one sees that, except for what might be called animal activities, such as ingestion, etc., etc., etc., men also perform actions which bear a characteristic peculiar to themselves, and these could be called ritualistic actions.
>
> But then it is nonsense for one to go on to say that the characteristic feature of *these* actions is the fact that they arise from faulty views about the physics of things. (Frazer does this when he says that magic is essentially false physics, or, as the case may be, false medicine, technology, etc.)
>
> Rather, the characteristic feature of ritualistic action is in no sense a view, an opinion, whether true or false, although an opinion—a belief—can itself be ritualistic or part of a rite.
>
> The religious actions, or the religious life, of the priest-king are no different in kind from any genuinely religious action of today, for example, a confession of sins.[20]

Wittgenstein, then, does not seek to criticize Frazer in any global or general manner for a supposed ethnocentric imposition of categories from another culture, and his position is not the culturalist or historicist view that stresses the incommensurability of different cultures taken as autonomous wholes. Rather, he complains that Frazer has compared the *wrong framework* of his own culture to that of another. Cross-cultural comparisons within the transcultural framework of "ceremonial action" are in fact encouraged by Wittgenstein, who does not hesitate to claim that there is an anthropological invariant called "genuinely religious action"—a basic point overlooked by any number of the exponents of a Wittgensteinian turn in the social sciences.[21]

[20] *Remarks on Frazer*, 67–68, 64.

[21] It is not necessary in the present context to draw up a list of the items in this vast literature. For a start, see Fred R. Dallmayr and Thomas A. McCarthy, eds., *Understanding and Social Inquiry* (Notre Dame, Ind.: University of Notre Dame Press, 1977), pt. 3: "The Wittgensteinian Reformulation," 137–218.

Wittgenstein's criticism of Frazer is based on both a negative and a positive definition of ritual. The negative definition denies that ceremonies and rituals are supposed to refer accurately to the entities they designate, and that they are meant to be effective or successful operations on natural reality. (Numerous examples of divinations, ordeals, cures, and curses in cultures all over the world are by the same stroke excluded from Wittgenstein's definition of ritual and mythology.[22]) The positive definition of the ceremonial variety of action is revealed in a passage that has been overlooked (and indeed censored) in the literature on these matters:

> When I am furious about something, I sometimes beat the ground or a tree with my walking stick. But I certainly do not believe that the ground is to blame or that my beating can help anything. "I am venting my anger." And all rites are of this kind. Such actions may be called Instinct-actions.—And an historical explanation, say, that I or my ancestors previously believed that beating the ground does help is shadow-boxing, for it is a superfluous assumption that explains *nothing*. The similarity of the action to an act of punishment is important, but nothing more than this similarity can be asserted.
>
> Once such a phenomenon is brought into connection with an instinct which I myself possess, this is precisely the explanation wished for; that is, the explanation which resolves this particular difficulty. And a further investigation about the history of my instinct moves on another track.[23]

The explanatory model proposed for "all rites" in this passage is a matter of what some contemporary psychologists would recognize as an instinct theory of motivation, whereby complex, symbolic behavior is explained as the consequence of so-called primary needs resulting from a physiological disequilibrium within the organism. The "instinct" to which Wittgenstein refers would be an inherited or innate psychophysical disposition capable of determining the perceptions and reactions of "something animal." It may be pointed out in

[22]It may be useful to document one example of the factual inaccuracy of Wittgenstein's notion of ritual. Wittgenstein mocks Frazer's idea that magical ceremonies could have been attempts to make a successful instrumental intervention in reality, and he notes, apparently in relation to ritual imprecations or malediction, that "Frazer would be capable of believing that a savage dies from error" (*Remarks on Frazer*, 68). Yet such cases have in fact been documented and represent an ontological scandal only for someone who holds that beliefs have no physical basis and could not have a causal role in biophysical processes; see Marcel Mauss, "Effet physique chez l'individu de l'idée de mort suggérée par la collectivité (Australie, Nouvelle-Zélande)," in *Sociologie et anthropologie* (Paris: Presses Universitaires de France, 1950), 311–30.

[23]*Remarks on Frazer*, 72.

this context that the major instinct theories of motivation have been subjected to extensive criticisms; although this kind of Darwin-inspired view of human cognition and behavior continues today in the branch of research known as human ethology, it seems fair to say that the kind of primitive version forwarded by Wittgenstein above was shown to be an erroneous approach to the explanation of human cognitive capacities as early as the instinct controversy of the 1920s.[24]

Wittgenstein's Dogmatism

It should by now be clear that there are problems with the substantive aspect of Wittgenstein's theory of cognition and forms of life. Yet this aspect is in a sense secondary to a more basic problem. That problem concerns the epistemological level, which has to do with what Wittgenstein can tell us about the way in which the forms of life can and cannot be known. What I want to bring forth in this regard is the essentially dogmatic nature of Wittgenstein's theory of knowledge. By this I mean to say that the ultimate answer to the question of what kind of methodological model Wittgenstein proposes is simple: there is none.[25] There is none, that is, if by "method" we mean the possibility, when confronted with a question or problem, of first considering alternative answers to it and second employing alternative means to the solution of that problem. Method in this sense is not a reified procedure, an algorithm, or thoughtless reflex whereby the answer can be produced infallibly. Nor is it a senseless craving for generality which leads to the obscuring of differences between particular cases. Rather, the word should evoke the possibility of reaching a position from which the observer may look

[24]For background, see K. B. Madsen, "Motivation," in *Handbook of General Psychology*, ed. Benjamin B. Wolman (Englewood Cliffs, N.J.: Prentice-Hall, 1973), 673–706. Inasmuch as the example concerns the violent "release" of anger, it is also pertinent to mention the extensive criticisms of instinctive and ethological theories of aggression and the alternative social learning model advanced by Albert Bandura in his *Aggression: A Social Learning Analysis* (Englewood Cliffs, N.J.: Prentice-Hall, 1973), and *Social Foundations of Thought and Action* (Englewood Cliffs, N.J.: Prentice-Hall, 1986).

[25]Bloor has criticized what he characterizes as the "methodological approach" to Wittgenstein, arguing that the real value of his work resides in what it can propose along the lines of a substantive theory of language games; see his *Wittgenstein*, 4–5. I am much less sanguine about the substantive rewards to be had from Wittgenstein's sociological moments and would point out that the substantive concepts of Bloor's book come from Mary Douglas, not Wittgenstein.

down and survey alternative "paths" so as to be able to choose one. That there should be reasons for this choice does not imply that the choice must be the right one, but that it is the best bet possible to the agent.

The kind of certainty Wittgenstein and his followers propose is based upon the rejection of methodology in this sense. The observer in Wittgenstein's *On Certainty* and in his remarks on *The Golden Bough* is anchored within a network of assumptions that are never the object of critical examination or choice. Rather, it is the inquiring agent's anchoredness in this "inherited background" which is supposed to make possible the indubitable status of certain basic beliefs and attitudes. Thus it is simply pointless for us to try to imagine that it could ever be otherwise, or that a more solid proof or justification could be given. Wittgenstein refers to this self-enclosed and autonomous background as a *Weltbild* or world picture, and also as a mythology.[26] He goes on to give examples of the kinds of truths it ensures, which range from simple arithmetical calculations and the laws of modern physics to knowing that one's hand is in front of one's face. The world picture is said to change over time; the stream flows along, the mud at the bottom is stirred up now and then, and underneath lies the riverbed, which shifts ever so slowly. The bank of the river, referred to as "hard rock," would seem to hold firm, yet Wittgenstein adds immediately that it could be that its movements and changes are simply imperceptible to those caught within the flux of shifting truths and falsehoods.[27]

Elaborated in this metaphor of the riverbed, and in the many Wittgensteinian texts that express similar conceptions, is an example of the kind of theory of knowledge I have earlier characterized as framework relativism (see pp. 23–24, 55–60). Just as the typical version of framework relativism insists that all knowledge is overdetermined by some kind of autonomous ensemble of governing conditions, so does Wittgenstein insist that the justification of knowledge is entirely framed within the mythology or world picture: "But I did not get my picture of the world by satisfying myself of its correctness; nor do I have it because I am satisfied of its correctness. No: it is the inherited background against which I distinguish between true and false."[28] More important, it is correct to speak of these Wittgensteinian views as a case of framework relativism be-

[26]*On Certainty*, §§ 95, 97.
[27]*On Certainty*, §§ 96, 97, 99.
[28]*On Certainty*, § 94.

cause we find here the same weaknesses typical of this prevalent type of position. To recapitulate briefly, framework relativism is upon close examination flawed by either its incoherent or its trivial character, and often both failings appear, so that there is a kind of global inconsistency as well. These doctrines are at a first level incoherent when they make two assertions: that beliefs are true or false relative to an inherited background, and that there are different backgrounds, such that for beliefs true relative to one background there is another background against which they are false. The central problem with such claims is the ambiguity and confusion inherent in the basic idea that a belief is true relative to a background. The clear, yet trivial way to construe this notion is that no subject believes a proposition to be true unless there is a background making this belief possible (which is a sensible statement about cognitive processes). Yet this is not a sensible or coherent statement about the ultimate validity or justification of the belief, and such a validity is essential to any correct definition of truth, which is a relation between a belief and a state of affairs external to the belief (and its cognitive background). Wittgenstein and the framework relativists in general are confused about the differences between such basic notions as (1) belief *x* is true, and (2) agent *S* with background *B* believes that *x* is true. The expression "true relative to a background" is a conflation of the two. To be warranted in asserting a case of the type (1), it is not enough that a sentence of type (2) be satisfied. When we go on to the second major clause of framework relativism, even more confusions are generated. On the framework relativists' own terms, it becomes possible to assert that the initial conflation of (1) and (2) is itself relative to some background—in some unspecified sense; and of course, the belief in the existence of radically divergent frameworks is also true, but only relative to some framework.

In short, framework relativism is trivial when it limits itself to statements about beliefs or cognitive processes but incoherent and muddled as a theory of truth. Yet it is committed to making statements about both and is thus globally incoherent. As such, this kind of theory of knowledge hardly offers a sound basis for rethinking the status of humanistic research.

But there is another important criticism of Wittgenstein's theory of knowledge. For in fact Wittgenstein does not offer us a "theory" of knowledge in any strict sense of the word; rather, what he proposes is the most rudimentary sketch of a very general position, a sketch that leaves many crucial issues unaddressed. One may very well wish to take seriously Wittgenstein's emphasis on the ways specific inquiries and beliefs are related to a far-reaching background of principles and assumptions, but many questions arise once this valid intuition is pressed and transformed into a series of strong theses about the determination of truth by something called "mythology." Some examples of these questions are the following. How, precisely, are beliefs conditioned or shaped by the "world picture"? By what specific processes is this conditioning achieved? Is it a matter of psychological realities, or something transcendental? Moreover, how is it that the world can effectively be organized into different and at times contradictory images? What must the world be like in order that it can be truly mapped by contradictory beliefs? How is it that *Weltbild wird Welt und Welt wird Weltbild?* How can a sentient agent ever come to know about the existence of the ground of beliefs operating "behind its back" to fashion its world? How could it know about the existence of other frameworks? These are some of the problems that have been in circulation ever since F. H. Jacobi took up the "paradox of the limit" in Kant; these are problems arising immediately in relation to any philosophy that begins to make strong claims about the complete overdetermination of truth by an a priori framework, and Wittgenstein's fragmentary comments do not really clarify them or present a detailed response.[29] Rather than assert baldly that conditions and backgrounds determine all beliefs, the question to ask in regard to scientific explanations is what conditions make it possible for beliefs to refer accurately—if always in an approximate and partial manner—to aspects of reality. Although it is true that the reliability of scientific method rests upon the logically, epistemologically, and historically contingent emergence of certain theories, this fact does not refute this reliability.

I have drawn a rather bleak picture of Wittgenstein's methodological model, one that may seem disproportionate in relation to his actual influence. I believe, however, that such an impression is erroneous, for Wittgenstein's conservative attitudes toward the role of philosophy, epistemological critique, and theory in the human and social sciences have been used to give a philosophical legitimation to an "everything is in order" trend in a variety of disciplines. It may be that this influence has had great consequences for the history of economics, for example, for it has recently been carefully argued by Olivier Favereau that Wittgenstein's later work may have had a

[29]Jacobi's complaint was that without the assumption of things in themselves it was impossible to get inside Kant's system of thought, yet with this assumption it was impossible to remain within the system. See F. H. Jacobi, *Werke*, vol. 2 (Leipzig: Fleischer, 1815), 304.

direct influence upon Keynes's softening of his "radical" criticism of classical economics. Favereau documents a major shift in Keynes's work, whereby the project of a major theoretical break with a classical model incapable of explaining involuntary and massive unemployment was replaced by a "pragmatic" tinkering with the classical paradigm. The second, moderate conception of the limited role of economic theory corresponds in many ways to Wittgenstein's strictures about the relation of theory to the forms of life.[30]

I conclude this discussion with a remark concerning the legacy of Wittgenstein within aesthetics and metacriticism. In a recent and highly perceptive article on this topic, Richard Shusterman claims that Wittgenstein's proper influence within these domains is not purely negative and antitheoretical.[31] On the contrary, Shusterman argues, for it is possible to present a few theses which amount to what Wittgenstein's work has to say about the nature of critical reasoning. Accordingly, these theses assert the radical indeterminacy of aesthetic concepts, the logical plurality of critical discourse, and the historical or diachronic plurality of aesthetic judgments. It seems that under such conditions, the only proof of the validity of a critical statement or judgment is its success, defined as the critic's ability to persuade others to look at the work in a way analogous to his or her own. Criticism can only be descriptive and persuasive, for it can neither formulate any general norms or laws nor demonstrate that its conclusions have any such bases.

These conclusions are familiar to anyone acquainted with North American literary circles, for some prominent critics have marketed this sort of Wittgensteinian line.[32] But Shusterman's account adds a few twists that do not figure in the standard sophistical version, which pits a purely "persuasive" criticism against a "demonstrative" science. Shusterman recognizes that, given the nature of the theses that have been laid down, the critic hardly has any basis for ruling out the possibility of deductive and inductive models of crit-

[30]Olivier Favereau, "L'Incertain dans la 'révolution keynesienne': L'hypothèse Wittgenstein," *Economies et Sociétés, Cahiers de l'ISMEA*, Série PE, No. 3 (1985), 29–72; and "La *Theorie Générale*: De l'économie conventionnelle à l'economie des conventions," *Cahiers Economiques d'Amiens*, forthcoming (Faculté de Droit et des Sciences Economiques, Université du Maine, 72017 Le Mans, France). I thank André Orléan for bringing this research to my attention, and I am grateful to Dr. Favereau for his comments on an earlier draft of this chapter.

[31]Richard Shusterman, "Wittgenstein and Critical Reasoning," *Philosophy and Phenomenological Research*, 47 (1986), 91–110.

[32]For background to my remark, see the excellent observations of Martha Nussbaum in her "Sophistry about Conventions," *New Literary History*, 17 (1985), 129–39.

ical reasoning. Although the Wittgensteinian view may offer an excellent description of some of literary criticism's efforts, "it is wrong as a theory of all critical reasoning."[33]

I wholly agree with Shusterman on this last point. I cannot, however, accept the naturalistic connotations of his final twist, in which he concludes his article by saying that, in the absence of rational norms about the nature of literary art and its evaluation, the fate of criticism is to be decided by the survival of the fittest: "Having identified and analysed the various critical games, he [the philosopher of criticism] must let them justify themselves, as they have justified and must justify themselves, in actual critical practice, in forms of life with art. Having distinguished between the different species, he must rely on the survival of the fittest."[34] Here, once more, and in a manner that is indeed typically Wittgensteinian, the abandoning of any real sociological analysis of symbolic practices only seems to clear the way for a return of nineteenth-century biological metaphors in which the specificity of social and institutional realities is obscured by pseudonatural imagery. The literary-critical marketplace has nothing to do with the survival of the fittest, and even if we were to try to apply a sophisticated evolutionary model to its analysis, this kind of Spencerian nonsense would have no place within it.[35] Although one of the virtues of a Wittgensteinian approach to aesthetics may be its insistence upon the role of the institution, which is an apparently sociological insight, it cannot be said that this insight has led to a constructive exchange between literary and sociological research.

Schon Immer Da, or the "Always Already" Argument

In the foregoing I have criticized some Wittgensteinian arguments against the possibility of a human science, arguments that hinge

[33]Shusterman, "Wittgenstein and Critical Reasoning," 109.

[34]Shusterman, "Wittgenstein and Critical Reasoning," 110.

[35]Literary critics interested in the possibility of an evolutionary approach to their own field are advised to make reference to the best available models of evolution in theoretical biology. Today this would involve paying attention to the various criticisms of neo-Darwinism's emphasis on fitness, an emphasis that has obscured the evolutionary role of such processes as pleiotropy, genetic drift, historical contingencies, group selection, and phenomena of self-organization and other intrinsic factors; Francisco Varela, lecture titled "Knowledge and Adaptation," delivered at the CREA, Ecole Polytechnique, January 1987. See also Elliott Sober, ed., *Conceptual Issues in Evolutionary Theory* (Cambridge, Mass.: M.I.T. Press, 1983); and Mae-Wan Ho and Peter T. Saunders, eds., *Beyond Neo-Darwinism* (London: Academic, 1984).

upon a questionable conception of science as well as upon some unsound ontological assumptions about the nature of human reality. Leaving this particular case behind, I now take up some other prominent statements on the same general topic, with the goal of isolating another major kind of antiscientific position. One type of argument in particular provides my focus. This argument, which I call the "always already" argument, is a philosophical motif or *topos* often employed in attempts to erect an epistemological barrier against scientific research into the origins of various cultural realities. I proceed by describing the type of argument in question through reference to a specific example; next I analyze this argument, seeking to reveal its key presuppositions; finally I criticize the use that has been made of this argument and distinguish between what I perceive as its legitimate and illegitimate applications. Inasmuch as many of the particular manifestations of this line of thinking have been directed at such notions as language, meaning, metaphor, and culture, the pertinence of these matters to literary research is great.

My example is taken from Jean-Luc Nancy's *L'Oubli de la philosophie*, a short essay inspired by the expressly didactic intent of setting forth a number of truths held to be crucial to post-Husserlian continental philosophy—which in Nancy's conception equals what is today genuine or valid in Western philosophy as a whole. The author claims that although the essay ultimately must be seen as articulating his own point of view, it is nonetheless his goal to set forth what is common to a certain (unspecified) group of contemporary thinkers, it being added that no one should obscure the very real differences between these thinkers' writings. We have here, then, a document claiming to be representative of an important intellectual tendency, easily identified as that of the contemporary French Heideggerians, postmodernists, or so-called poststructuralists. Although the point is by no means essential to what follows, I think that Nancy is right in viewing his text as exemplary, for its views are indeed taken as gospel by a number of influential thinkers; but I disagree with the idea that these attitudes in fact constitute the truth of real or genuine philosophy on the whole, and indeed I find that an extremely narrow perspective. Such is the context of my first example of the "always already" argument. The reader unfamiliar with Nancy's text is asked to keep in mind that throughout his essay Nancy develops his own idiosyncratic distinction between *sens* and *signification*, terms I translate as "meaning" and "signification," respectively. I have tried to respect the singularities of the style of the French original, which does not always make for the clearest English prose:

> A signifying world is a world offered to understanding, explication, and interpretation before it has any signification. Our world is a world presented as a world of meaning, both this side of and beyond all constituted meaning—for example, on this side of and beyond the meanings of the words "world" and "man." This presentation of its meaning or in meaning, this *elementariness of meaning*, takes the place, in a way, of the schematism. But unlike the schematism, it does not perform an operation, is not a "hidden art," and is the condition of neither the possibility nor the production of significations: if it is our most specific possibility, this is in the sense that we are "capable of meaning" (*nous pouvons le sens*), and we are capable of it without conditions. Nor is the elementariness of meaning a primitive origin of significations, no more so than signifying words are made up out of signifying phonemes, as Plato already noted. There is no meaningful provenance of meaning—nor, of course, is there a meaningless provenance, which would still presuppose meaning. Quite simply, *there is no provenance of meaning*: it is presented and that is all. This is why meaning, if it is an element more hidden than anything a logic of elements could specify, is also something perfectly unconcealed, offered immediately with our existence.[36]

In drawing a broad opposition between "meaning" and "signification," Nancy establishes a series of priorities: on the one hand, he situates any number of outdated and illusory ideologies and systems of thought which he characterizes as "significations"; on the other hand, he presents his thesis of the elementary character of meaning, which he describes as the basic and first level or element of our existence, thought, and experience. The modern significations are for Nancy part of the project of a Western metaphysics that must now be put in question, and to which there have emerged real alternatives; this questioning or "re-orientation" of Western philosophy in fact turns on the central notion of a "project." According to Nancy, the most basic framework of Western metaphysics was the kind of desire that animates a project. The project implies a two-fold schema: there is a fundamental "lack," and then a corresponding quest for a past or future signification that the project, when completed, will realize, thereby remedying the lack. Meaning, then, is not properly speaking to be understood as the endpoint of any project, whereas significations are the illusory goals and norms entertained by the subject who still believes in projects and is caught up in this particular formation of desire. Meaning is not to be the goal of any interpretive or representational quest, it is not the fruit of any

[36]Nancy, *L'Oubli de la philosophie* (Paris: Galilée, 1986), 91.

philosophical, scientific, or practical endeavor; it is always already there. It follows from this thesis that there cannot be some past origin or conditions of possibility which must be fathomed if meaning is to be properly understood. Thus the "always already" argument is in this context linked to the defense of Nancy's contentions about the elementary or immediate nature of meaning within the sphere of human experience. Indeed, Nancy goes so far as to say that meaning is prior to humanness.

Following these reasonings, then, it is illegitimate or wrongheaded to undertake an inquiry into the origin of meaning. Thus, one of the results of Nancy's assertions is that a certain question is ruled out, having been judged in advance as lacking any valid answer. Given the import of the question, it is clear that the validity of this epistemological prohibition would have grave consequences for the project of a human science (and indeed Nancy does not hesitate to include science in the list of forlorn projects). To understand Nancy's argument, we must grasp the precise nature of the question he deems illegitimate and isolate the assumptions that could lead someone to arrive at such a conclusion. I shall try to reconstruct these points in three steps.

1. We may begin with the pair of phrases "meaning" and "the origin of meaning." For many, it goes without saying that if meaning is a phenomenon or group of phenomena worthy of scientific investigation, an inquiry into the proximate origins or conditions of the emergence of this phenomena is *in principle* legitimate. It could very well be decided that the term lacks precision and that clarification must be provided whereby more specifically delimited realities associated with this term are isolated for study.[37] Yet, in any case, the initial assumption is that it is possible and legitimate to inquire into the anterior conditions of the emergence of meaning. One asks, then, what conditions make meaning possible, what must be there for it to be a reality (if it is one), for what reasons it is the way it is, how and why all this comes about, and so on. The assumption whereby these and similar questions are deemed legitimate or coherent does not entail another assumption, however, namely that science must in principle be capable of giving correct answers to these questions. Even less does it mean that we imagine that one day we must certainly possess perfect and definitive

[37]A specific example related to this very issue is presented in Charles Morris's discussion of vagueness and confusion surrounding the term "meaning"; Charles Morris, "Foundations of the Theory of Signs," in *Foundations of the Unity of Science: Toward an International Encyclopedia of Unified Science* (Chicago: University of Chicago Press, 1938), vol. 1, no. 2; 1–59, esp. 43–48.

responses to these queries. It could be that no one will never discover the real nature of these conditions and processes. But it *is* assumed that there is no a priori argument that can tell us in advance that the research must fail. Yet this is precisely the type of opinion that Nancy expresses in his "always already" argument. How are such conclusions reached?

2. Nancy poses the following question: does the origin of meaning itself have meaning? This is a crucial step, and it does not go without saying. My central criticism of the "always already" argument focuses on the issue of the appropriateness of this question. But first it is necessary to retrace the answer given to it.
3. One of Nancy's conclusions is that there is no origin of meaning. He proclaims that the "presentation" of meaning is an elementary and seemingly inexplicable fact: meaning is always already there. This elementarity of meaning is not an origin, however, for meaning is said to have neither a meaningful nor a meaningless provenance. Why? If the origin of meaning had meaning, it would be necessary to admit that this origin was not really the absolute origin of meaning, because meaning would have existed already; a meaningful origin of meaning would imply that there was meaning before the origin of meaning. But what happens if we assert the contrary, saying that the origin of meaning is meaningless? The objection here seems to be that the very idea of something being meaningless is only coherent if meaning is possible, which is not the case before all meaning. Thus the origin of meaning is neither meaningful nor meaningless, because both of these assertions imply that the opposition meaningful–meaningless transcends its own origin and can be applied to that origin. Given this conclusion, Nancy goes on to assume that the dichotomy meaningful–meaningless is exhaustive and hence applies to whatever there is. There can be no question of getting behind, above, or beyond this dichotomy; there is no source where it would not yet be pertinent. It follows that there is no origin of meaning, that meaning is "elementary" and there "for us"—"before any anthropology."[38]

My criticisms of Nancy's assertions have to do particularly with the second moment identified above, in which the question was raised whether the origin of meaning was itself meaningful. What is at stake in such a question is how we conceive of the relation between a phenomenon or state of affairs and what is called its origin. Given Nancy's assumption, the question he asks is at once legitimate and necessary; but what is this assumption? It can be formulated in the following manner: given a phenomenon *x* and its origin,

[38]Nancy, *L'Oubli de la philosophie*, 90.

there must be an absolute transcendence of the origin in relation to *x*. The word "origin" thereby means something like "transcendental condition of possibility," where this latter refers to a strict hierarchical difference of level, an absolute priority, in the relation between the conditions and the conditioned. Such conditions must be totally prior and external to the domain or state of affairs they condition; thus an origin of meaning having meaning could not represent a true origin in this sense, because such an origin would already have a property of which it is said to be the origin. Is meaning itself, then, to be taken as the transcendental condition of the possibility of experience? One is tempted to think so, for Nancy speaks at one point of an "eternal sovereignty of meaning."[39] But he also insists upon the perfect *immanence* of meaning, which he declares is *not a primitive origin*, and which he says is not the condition of either the possibility or production of signification.

To appreciate what Nancy is doing, we need to recall briefly the role a similar kind of argument has played within the history of philosophy. All that needs to be remembered is that, in general, whenever philosophers have declared that some state of affairs is the case always and already, what they were usually seeking was a source of certainty which could be relied upon in refuting some other set of contentions. Thus the philosopher responds to the skeptic who is trying to undermine confidence in abduction as a whole that the cogency of the skeptic's own arguments relies upon this very principle, which is always already there. What we can be said to think "always and already" has a universal character, then, and thus may be deemed a source of certainty, grounding, and legitimacy. These are, of course, precisely the notions such postmodern thinkers as Nancy now reject as vain and illusory ideals. Thus if the form of the argument is similar, the specific intent has been displaced, for it is now a matter of establishing the amorphous and plural term "meaning" in the place of more specific, anchoring "transcendental conditions." Nancy, then, can be read as taking up the traditional philosophical quest for an absolute origin and foundation, but this quest has been displaced from a positive to a negative theology, from the search for a source and starting point to the discovery of the absence of origin. This absence is taken as the supreme and sovereign "presentation," which is posited as the absolute and inexplicable condition behind our experience and knowing.

We can hardly fail to recognize here the reworking of some Heideg-

[39]Nancy, *L'Oubli de la philosophie*, 103.

gerian motifs. Indeed, Nancy's argument about the origin of meaning rehearses some of the key movements of the argument over Being in the beginning of *Being and Time*, including the essential attempt on the part of the philosopher to usurp the "positive sciences." With the early Heidegger the long-standing philosophical motif of "always already" becomes a way of asserting the priority of his own *Seinsfrage* over so-called positive research, for Heidegger states that problematic ontological foundations are always there already, lurking behind the back of sciences that simply take them as self-evident.[40] And along with the priority of the question of Being comes that of the *Daseinsfrage*, presumably because *Dasein* is the being that asks the primary question of Being, although it should be noted how little argument Heidegger gives for this pivotal claim. Yet it is taken for granted that the "analytic" of *Dasein* is by the same stroke situated prior to research in anthropology, psychology, and biology. (The claim about the latter is particularly curious, for it is flatly stated in the same context that biology is "founded upon the ontology of Dasein, even if not entirely.") With Nancy and others, similar gestures are made to assert the primacy or priority of meaning, or of language, or writing, or representation, or the sign, or any number of other concepts and neologisms. Similarly, Nancy's prohibition against inquiry into the origin of meaning echoes what Heidegger proclaims about one way the question of Being should not be treated: our first philosophical step cannot be a matter of telling a story, of "defining entities as entities by tracing them back in their origin to some other entities, as if Being had the character of some possible entity."[41] When Nancy asks whether the origin of meaning itself has a meaning, he echoes Heidegger's rhetorical question concerning whether the Being of entities could itself be an entity, and in both cases the response is a matter of asserting some essential difference—in my terms, a hierarchical gap between the conditions and the conditioned. Thus Heidegger asserts in the same passage that "the Being of entities 'is' not itself an entity," and we know by now that the presentation of meaning is not just another meaning. Not only must we insist upon the priority of the question of ontological foundations and absolute, unconditioned conditions, but we must also respect the sheer impossibility of ever being able to speak positively about any such things, for the subjects of any such inquiry are always already on the other side of the hierarchy.

[40]Martin Heidegger, *Being and Time*, trans. John Macquarrie and Edward Robinson (New York: Harper & Row, 1962), 75.
[41]Heidegger, *Being and Time*, 26.

My criticisms of such assertions will be brief. I do not intend to try to settle the question of the validity of the argument concerning the totality of beings, or Being—which is not to imply that such questions could ever be settled. It makes some sense to me to assert that neither the attributes of "being" nor of "non-being" should be applied to the origin of all beings or entities. And I see no room for positive inquiry into the age-old metaphysical question, Why is there something rather than nothing? Although it may seem right to say that everything *within* the universe either is or is not, somehow Hamlet's question does not seem applicable to the origin of it all, and my own theology would be to say that the universe *is* always already. But it does not follow from this sort of speculation that we are right in applying similar arguments to any question whatever; on the contrary, we are warranted to ask whether such an application is justified. How was it determined that concepts and realities related to culture or the anthropological sciences should be treated in the same manner as the concept of the totality of beings, or Being? Is human culture as a whole an autonomous, self-enclosed Totality? How was it determined that nature is only one area of being, alongside which there are others such as *Dasein* or language? Did Heidegger really prove that the question of *Dasein* comes before any psychology or anthropology, and "certainly" before any biology?[42] Whether it comes before them in his work is not the question, and thus an answer cannot be supplied merely by means of further Heidegger commentary. My point is that Heidegger in fact provides no detailed arguments for these crucial decisions. Such beliefs are simply asserted by philosophers who do not want to see human and cultural realities ranked as "entities among other entities," for then the autonomy and priority of their own negative-ontological discourse vanishes, giving way to the theoretical and explanatory efforts of the physical, biological, and human sciences. But what matters are the consequences, not the motives, of such positions. By insisting on the problem of absolute origin, Nancy and others overlook, and indeed actively discount, any number of other kinds of questions having to do with origins. "Origin" in this second sense is not a manner of "transcendental conditions of possibility," of absolute priority and exteriority. The term "origin" can also be used to refer to the anterior conditions of a state of affairs—or again, to the proximate causes of *x*. I have seen no valid arguments as to why it is impossible to conduct research into this latter type of condition in

[42]Heidegger, *Being and Time*, 71.

regard to such anthropological realities as meaning. The sciences in fact have a great deal to say, for example, about the physical, biological, and neurophysiological conditions that must be satisfied before there can be any human experience of meaning—which is not to say that the origin of life on this planet is not still veiled in mystery, or that we must someday possess a definitive account of the emergence of human culture. But unless we adopt a theology asserting the absolute priority of the human as such, we must realize that there is no reason to believe that human culture, or language, or experience was always already there and thus had no originating conditions (as one may well wish to say of the Totality).

Here we touch upon the question of reflexivity, which sometimes is thought to support similar arguments against the possibility of a human science. The thrust of this type of contention is that a human science would require the agent of such knowledge to be fully aware of the conditions behind his or her own knowledge, which is taken to involve an impossible regress insofar as the presence of the agent is one such condition and full self-reference is impossible. But this argument is only applicable when the self-knowledge in question is defined as total knowledge, and it is extreme and unnecessary to work with such a definition. The realization of particular explanations of specific aspects of human realities does not require the total "self-reflection" of some global pseudosubject called "man," and thus the full-fledged reflexivity argument imposes an unrealistic and irrelevant requirement on human science. Echoes of this kind of argument, however, appear quite frequently. Again and again it is taken for granted that *because* knowledge is made possible by conditions of which the subject may not be aware, this knowledge is somehow invalidated—particularly when this knowledge is supposed to refer to human realities (of which the subject is "of course" a part), and even more so when what is to be explained are the conditions of cognition, meaning, reference, and so on (for clearly the knower is "conditioned" by them). But these kinds of clauses do not entail the infinite regress of a full reflexivity, and to think that they involve some major obstacle to knowledge is a mistake. Inasmuch as this mistake is particularly widespread, some additional comments are necessary.

It is true that the agents of any inquiry into the conditions of the emergence of human realities are themselves part of this reality in a broad sense, but this apparent identity of the knower and the known vanishes whenever the objects of the inquiry are given a more precise delimitation. The anthropologist is surely a member of a culture, but not necessarily of the one under study. But are not the anthropologist's theories and findings still conditioned by conditions significantly similar to those being studied, insofar as such inquiries came after the emergence of culture and rely upon its many fruits—such as language, developed cognitive processes, cultural traditions and practices? Surely in the absence of such conditions there could be no anthropological knowledge. Indeed, but what follows from this intuition? Does the fact that anthropological knowledge has anthropological conditions entail the impossibility of its truth? Not at all. It is important to recognize that scientific explanations are only possible under certain kinds of conditions, but an absolute disjunction of the domain of the knower and the domain of the known is not one of them. Physicists explain physical systems in which they are located; biologists are alive, but they can explain aspects of living organisms, including human ones; historians, who are "in" history, may nonetheless be able to explain aspects of certain past events. These examples of explanation are possible because many different conditions are satisfied, and it is mistaken to try to rule them out a priori on the basis of a supposed reflexivity. Yet in literary theoretical circles it is frequently reasoned that, because it is only possible to conduct an inquiry into language and meaning with a language that is already meaningful, such an inquiry is a vicious form of circularity.

To evoke a more specific example, one may point to Jacques Derrida's insistence that Aristotle's philosophical attempt to define metaphor necessarily fails because the definition is conditioned by a language that is itself inherently metaphorical; the *physis* that is supposed to be the natural origin and ground of a human *mimesis* can only be designated within *mimesis* itself.[43] Indeed, the failure of this and other attempts in the history of philosophy is said to represent the necessary failure and death of philosophy's perennial effort to ground itself. The *philosopheme*, which is presented as standing outside the tropic history of oppositions between what is figurative and literal, metaphorical and proper, is in fact not an "originary" and "fundamental" term, but another trope. Moreover, in a particularly valuable passage, Derrida informs us what would be required to avoid this failure of the *philosopheme*: "It would be necessary to assert that the meaning aimed at through these figures is an essence, rigorously independent of what carries it, which is already a philosophical *thesis*; one could even say that it is the *single thesis* of philosophy,

[43]Jacques Derrida, "La Mythologie blanche," in *Marges de la philosophie* (Paris: Minuit, 1972), 247–324, esp. 301–2.

that which constitutes the concept of metaphor, the opposition between proper and non-proper, essence and accident, intuition and discourse, thought and language, the intelligible and the sensible, etc."[44]

These are far-reaching and provocative contentions, accompanied by some intricate readings of a variety of philosophical discussions of metaphor. But these intricacies do not attenuate the baldness of the main point being asserted here. The truth of this point depends on what is meant by the notion of "rigorous independence," which is evoked in regard to the relation between a figure and its meaning, on the one hand, and the essence to which it refers, on the other. If by "essence" is meant "extradiscursive reality not owing its *being* to the act of its designation by a sentient agent," then we would indeed agree that this is *the* thesis of philosophy, but also of science and most of our ordinary efforts to assert a truth about states of affairs in reality. In other words, we frequently assume the possibility that reference gives us epistemic access to recurrent patterns and properties of particular, mind-independent realities, which we may or may not wish to call "essences." We have many reasons to believe that we are quite frequently justified in holding this thesis to be valid, and it is an error to suppose that the philosophers whom Derrida analyzes have any monopoly on this thesis, or that its validity could be determined by sole "reference" to their writings. But of course the "rigorous independence" in Derrida's phrase could also be construed in some severe and unusual manner, as in "bearing absolutely no relation to each other," "sharing no conditions whatsoever," or "absolutely disjoined from each other." Then the fact that the knower is at one level or another necessarily situated within the same realm as the known (for example, the fact that philosophical writings about language are themselves part of language; or the fact that historians are "in" history) could be pointed to as evidence of a circularity invalidating the thesis. But such demonstrations merely lead us back to the incoherent and irrelevant forms of skepticism and tell us nothing specific about the possibilities of the human sciences. If true and justified belief is beyond our grasp, I see no reason why those who write volumes announcing this human failing should choose to single out the weaknesses of the history of philosophy, linguistics, or the human and social sciences. Is there any reason why they do not focus on the contemporary natural sciences?

[44]Derrida, "La Mythologie blanche," 273.

Reductionist Materialism and the Elimination of the Human Sciences

As I stated earlier, it is important to note that objections to the possibility of the human sciences do not come only from the side of the humanities. A brief illustration of this point may be useful, for it is probably the case that many literary scholars ignore this fact in their concern with the dangers of scientism.

Let us imagine, then, a rejection of the possibility of the human sciences motivated, not by an insistence on the unique and inexplicable qualities of such human realities as value, freedom, or aesthetic experience, but by a wholesale attack on the notion of intentionality implied by all talk of such realities. It would be argued, for example, that to speak of intentional action and of mental states is a wholly inaccurate way of referring to what are in fact material processes obeying physical laws. Let us suppose that we grant this rejection of the entire language of intentionality. We also refrain from asking the usual questions about the reductionist's own necessary reliance upon this language. What we want to explore are the implications for the possibility of a human science. The consequences are indeed immense, for human actions are reinterpreted as mere events that, even if they can in some sense be associated with the movements of human beings, are not characterized by any specifically human properties. Insofar as the human sciences are concerned with the illusory entities postulated by the language of intentionality, they are engaged in a futile attempt to describe and explain actions, beliefs, values, persons, and so on, and not the underlying forces and processes that are the true causes and ultimate realities. The reduction of *actus humani* to *actus homini* implies that the task of explaining our so-called acts and thoughts can no longer be taken up by the "human sciences," which, because they have been revealed to have no real object of investigation, can be eliminated. Explanations of "the human" can be provided by a materialist account of the physical and biological processes responsible for sentience: mental states are *defined* as physical processes. And let us note before going on that this sort of approach seems to be supported by a lot of well-confirmed beliefs, including those of the advanced physical and life sciences, as well as those of the manifest or everyday image of reality (e.g., the many apparently causal correlations between such physical events as taking drugs and ensuing mental states).

Does this full-fledged extension of the natural science's explanatory program result in the elimination of the human sciences, then? Is this the ultimate meaning of the unification of the sciences? At this point we seem to have come full circle; we first discussed arguments against human science animated by rejections of positivist unifications, now we encounter a line of thought whereby the project of the human sciences is written off as a result of the natural science's own zealous project of reductionist unification. The relation between the two antithetical refusals of human science is probably not purely contingent, and we may speculate that many of the humanistic arguments against the possibility of scientific explanations of "the human" are motivated by a deep-seated rejection of the reductionist program, which is thought to be entailed by the scientific model as such. Such motives are wholly understandable, but the conclusion is nonetheless mistaken. In order to avoid the reductionist destruction of the anthropological disciplines, one need not mount arguments against science as a whole.

It is not my intention here to present a detailed solution to the tremendously complex and difficult issue known as the mind–body problem. All that needs to be said is that at the present time it is possible to defend a well-argued position that obeys the crucial constraints of a general, scientific materialist conception, including those constraints set by many specific doctrines and findings of neurophysiological research, without entailing the reductive view whereby mental phenomena are defined uniquely in terms of the laws of physics. In other words, upon this view mental states are indeed taken to be states of the central nervous system, as the materialist would have it, but this contingent fact amounts neither to an exhaustive identification of the mental states nor to a specification of their ultimate essence. The identity of the mental and the physical, then, is held to be contingent in the sense that H_2O is identical to water. Thus the identity theses of materialism are not rejected as a whole, which would have the result of returning us to epiphenomenalist or dualist views. Yet the functionalist position modifies the sense of this identity by ruling out all reductionist physical definitions of mental processes and states.[45] But how does this position save the human sciences? It does so by blocking the first step whereby the language of intentionality as a whole is discarded in favor of

[45]See Richard Boyd, "Materialism without Reductionism: What Physicalism Does Not Entail," in *Readings in the Philosophy of Psychology*, 2 vols., ed. Ned Block (Cambridge, Mass.: Harvard University Press, 1980), 1:67–106.

doctrines describing the underlying physical, chemical, and biochemical processes. Instead, the functional account holds the way open for the partial autonomy of the mental, not by postulating the autonomy of wholly nonphysical capacities or acts (as in idealism), but by conceiving of an intentional causality that would be realized in, and could act upon, physical states. Insofar as beliefs, decisions, and judgments would figure among these mental states, the epistemological bases of the human sciences, as well as their objects, could in principle be rescued from the reductionist challenge.[46]

The Interest of a Human Science

Early in this chapter, I suggested that one should distinguish between the question of the possibility of the human sciences and the related but distinct issue of their desirability. This distinction tidies up things nicely, for one can thereby keep the purely epistemological possibility of a certain type of explanation separate from various practical and political issues pertaining to the role of science within social life. This distinction would then be read out at both the individual and institutional levels of analysis. For example, those who argue for the possibility of a scientific explanation of cultural realities are not committed to the belief that such an explanation is always the best form of cognitive stance to adopt, nor must they believe that this stance captures the totality of human experience. Only scientism contends that scientific explanations are the only form of truth, and it is of course ridiculous as well to hold that truth is the only good. The social interests and consequences of human science can, then, be isolated as an issue which, if it is of great importance, remains in significant ways distinct from the question of scientific truth.

But in spite of its partial validity, the broad distinction just defended in fact amounts to an oversimplication that mirrors the positivist disjunction of normative and factual judgments. As I have

[46]See, for example, Myles Brand's argument that action theory can remain neutral on the status of mental events and their specific relations to the realities of the brain; there is no need to resolve the outstanding issues on the theory of the mental. This is possible, of course, only as long as we are not required to accept the conclusions of a radical eliminative materialism; *Intending and Acting: Toward a Naturalized Action Theory* (Cambridge, Mass.: M.I.T. Press, 1984), 44–45. For a functionalist defense of the (partial) autonomy of levels of description, see Jerry A. Fodor, "Special Sciences, or the Disunity of Science as a Working Hypothesis," *Readings in the Philosophy of Psychology*, ed. N. Block, 1:120–33.

suggested earlier in this work, the so-called value neutrality of science is a myth for at least two main reasons: (a) because the goal of systematized knowledge animating scientific inquiry itself entails a normative choice in favor of justified knowledge and against dogmatism, ignorance, mystification, falsehood, and so on; and (b) because, within this broad normative choice in favor of a certain understanding of knowledge, actual scientific inquiry must be guided by additional *local* decisions concerning the direction and goals of research (unless one remains purely within speculative philosophy, it is impossible to do research on "man" or "the social")—on the contrary, specific contexts and aspects must be selected for examination, with particular questions and modes of explanation in view.

Although the implications of these points for the natural sciences are complex and important, I limit the present discussion to a brief and highly schematic consideration of their bearing upon dualist arguments against the possibility of the human sciences. There are essentially two main questions to be raised in the present context: (1) is there any way to ground the scholar's essential orientation toward knowledge? and (2) is there any good argument to the effect that the value orientation expressed in (a) above is somehow inapplicable in relation to the anthropological sphere? The two may be taken up in order, and in respect to each the answer follows the same logic.

I shall not try to argue in detail here for the superiority of an orientation toward truth on the part of researchers within the human sciences. In fact there would be something absurd about making such an argument in the present context. Briefly, I think that such a value orientation follows from the most basic institutional norms governing research practices, and that one cannot coherently occupy the institutional role of researcher without assuming the validity of that orientation in relation to these institutional activities. (I am begging the Jean Genet/Michel Foucault question of whether literary scholarship is in fact a cult or a brothel and hence does not know any such institutional norms, just as I am begging the question of whether some future research institutions should not discard the goal of knowledge.) Given this kind of institutional framework, there is no problem of having to ground the normative judgments in favor of knowledge. But this grounding remains relative to a prior fact, and it does not follow that this particular constellation of priorities and evaluations is in general to be elevated over others. I am certainly not saying that this knowledge orientation is the way to the good life or should be embraced because it is the path to human progress, nor am I even going to argue why the community of literary scholars should not be a cult. As is evident throughout this book, I have notions on these topics, but this is not the place to set them forth. On this score one may refer to Habermas's efforts to yoke the will to truth to the will to understanding, emancipation, justice, and equality, but it should be pointed out that one needs heavy machinery to bind human history and its determinations into a teleological organism headed necessarily toward growth—and not only toward growth, but toward good growth. Peirce was able to invest in this kind of eschatological engineering only at the cost of wild cosmological speculations—his "synechism" in which the universe is fancied as being destined toward increasing order and consciousness (*Welt wird Geist*). But it is dubious that such speculations can be taken literally today, and it is really not clear that they receive any genuine basis in Habermas's references to Piagetian genetic structuralism (a Hegelianism of the individual organism's development), *and* Chomsky-style depth competence structures—even if we add into the mixture a systems theory that stands in for "the ruse of reason" and the "laws" of dialectical materialism. Here we have a whole checkbook's worth of blank checks, incapable of paying off on the one crucial question: what if the "fixed point" of the self-organizing system called human history does not correspond to any of the "subjective utilities" of the players in the game, and in particular, those of the philosopher? There is at the present time no guarantee, scientific or other, that this is not the case. But that is no warrant for nihilism.

What of the second question, which concerns the implications of normative matters for dualist arguments? Is there any more fuel to be had for them here? The discussion can be fairly brief, as I think it is easily shown that the *general* topic of normativity adds nothing to the battery of skeptical and dualist arguments. Consequently, attempts to stress this additional issue only add more tangles to positions that are already wholly contorted. What would it mean to contend that human beings are incapable, in a global and a priori manner, of ever effectively orienting their actions in terms of the goal of arriving at true and justified beliefs concerning human realities? One could of course put in question the particular will to truth along with all forms of rational or purposive action, and we would quickly arrive at various forms of skepticism, coherent and incoherent. But these stances are not what is in question here, and I take for granted minimal assumptions inherent in the schemas of basic intentional action (e.g., assumptions about actions being oriented by action plans and serving as causes). Only on this assumption may we pose the specific issue of whether there is any good reason to hold it impossible that action could be oriented toward truth, justified knowledge, maximal epistemic utilities, and so on. One kind of argument that is typically presented in this regard has it that other kinds of interests prevent scholars and scientists from achieving objectivity. Personal and political interests, the generally ideological nature of thought, the will to power, and so on, would be obtacles to any genuine orientation toward truth, while these very same factors would motivate all of those actions that masquerade as the desire for truth. It would of course be absurd to deny that these and other interests have frequently distorted attempts to explain sociohistorical realities, and it should also be conceded that when research is conducted under certain institutional conditions (e.g., in cults and brothels), one has every reason to expect the intents as well as the findings to be turned away from the truth. Yet this valid insight does overlooks the fact that some of the other interests that may motivate the formation of beliefs about social realities are at the same time necessarily linked to the goal of truth, which they may advance, at least in some approximate and partial way, in spite of other priorities.

These are significant issues, but it should be recognized right away that they can only be approached correctly in a local and particular manner, for any truth-seeking inquiry into them presupposes in principle the possibility that social research need not always be distorted by other interests. In other words, this rather minimal norm is always already there once one poses the question of inquiry, and thus its global negator engages in a performative self-contradiction productive only of undecidabilities. So either the issue is undecidable, or we can undertake knowledge projects.

Thus there is reason to doubt the consistency of supposedly groundless or decisionistic-fatalist rejections of the will to truth, and this kind of literature indeed reveals a thicket of twists, turns, and oscillations. One observes, first, the extent to which the antiknowledge impulses fail to maintain an absolute separation from the will to truth. Claiming to have any knowledge whatsover of the conditions and consequences of science, for example, reveals that one holds certain implicit assumptions in favor of the possibility of social knowledge. As a result, there are problems for those who try to base their opposition to social science uniquely on moral, pragmatic, or political grounds. They have difficulty justifying their various assertions about factual social realities while maintaining rigorous skepticism about knowledge. They must tell us why we should believe that a scientific attitude toward human affairs necessarily produces bad effects (by means of a "scientization of politics," for example). Yet they have supposedly already foresworn all such reasonings. Oddly, such writers never say that social science must fail simply because they do not happen to want it to succeed. Nor do they assert that there simply are no reasons for this particular preference. Instead, they feign to give us reasons, and in doing so end up relying on various concepts of social and psychological reality, just as they make arguments that cite facts and figures and include lengthy references to the literature on the subject. Why all of these footnotes? we want to ask them; and if the answer is that this is part of the apparatus used in this particular game, we may still inquire as to why they are playing this game at all. Why have they not "passed by"? Of course the various acts of epistemological faith we may in this way discover in their discourse are no evidence of any credence in scientific knowledge in any rigorous sense. But these arguments are all the same situated within a normative context defined by basic explanatory patterns and principles which are ultimately those of science. Perhaps these learned antiscientific arguments are sheer hypocrisy, in which case the pragmatists and antirealists really do not believe in the goal of truth, but it is decisive that they have to pretend to play by the rules of the game of knowledge whenever they want to argue against the whole enterprise. In short, it hardly seems compelling to mount a moral or political rejection of social knowledge if one cannot defend one's own rejection of it without a performative self-contradiction. It should be noted, finally, that the argument I have just proposed takes the same form as those used by Apel and Habermas in their foundational projects, yet I would defend only an empirical, not a transcendental version. But at this point in the present text, argumentative economy has been sacrificed to emphasis, and it is best to move on.

PART FOUR

LITERARY KNOWLEDGE

In keeping with my belief that the role literary theory should play at the present time is primarily epistemological and comparative, the following analyses of literary knowledge are not meant to provide a substantive theory of literature—by which I mean a systematic, rigorously organized body of propositions concerning a well-delimited object domain called literature. Nor am I proposing a prolegomena to such a theory of literature, not only because such programmatic gestures can rarely be followed up in any consistent and detailed fashion, but because it is simply not obvious that the correct goals of literary knowledge lie in this direction at all. As I announced from the very beginning of this book, it is not my opinion that literary knowledge does not really exist yet, or that what criticism needs is a completely new starting point, an absolute foundation upon which a truly methodical and systematic theory could be built. But at the same time I totally reject the notion that everything is in order in the literary critical institution exactly as it stands, and that to call for major changes is to be a traitor to the profession. One should be wary of anyone who asks us to choose between the Tower of Babel and the Crystal Palace, and it strikes me that the "new professionalism" remains inside that same useless opposition. Literary theory, which is in any case a hypothetical, exploratory, and fallible endeavor, should attempt to elucidate both what is strong *and* weak within literary culture. Ideally, theory would contribute only to the former. Yet an active criticism of the latter may be a necessary and crucial part of that contribution. The last thing we need is a critical theory that calls for business as usual.

The notion of literary knowledge is, of course, so broad as to be extremely vague. Is it knowledge *of* literature or knowledge *in* literature? In what sense is this knowledge literary? Is anything specific at all designated by these two terms? Not only is the expression vague, but it is, to a certain extent, unorthodox, for our various traditions have often tended to disjoin its two terms: literature is illusion, fiction, mere appearance, the opposite of real knowledge. Or literature is protected from this accusation by being sealed off from the sphere of real cognition and truth values: as the phrase goes, "Now, for the poet, he nothing affirms, and therefore never lieth."[1] Pseudo-statements may be healthy, erotic, invigorating, but this does not make them knowledge.

Critics should be able to argue for the cognitive value of literary knowledge in both major senses; that is, it should be possible to demonstrate the possibility and value of a knowledge of literature, as well as a knowledge in literature. Although many people—and not only literary critics—are prepared to grant both points without a moment's hesitation, as soon as one starts to say anything more explicit about the nature and content of such knowledge, enormous difficulties become apparent. One may experience this in conversation with colleagues from other disciplines, who are usually quick to agree to the cognitive value of literature and its criticism, and who are thus only puzzled by the specialists' difficulties in saying anything clear and rigorous about this knowledge. At such moments one understands the appeal of purely negative definitions of the literary insight, as in Maurice Blanchot's *L'Espace littéraire,* where the privileged figure is an Orpheus who must never turn to embrace his Truth, for in that very instant she vanishes into oblivion.[2] Yet to say even this is perhaps already to have said too much, for this kind of literary knowledge remains forever shadowy.

In Chapter 6 I begin with a discussion of the dominant, institutionalized forms of literary epistemology, that is, with the main trends of what is known as critical theory in North American circles. Although this literature is large and diverse, it is useful to isolate

[1]Sir Philip Sidney, "An Apology for Poetry," in *Critical Theory since Plato,* ed. Hazard Adams (New York: Harcourt Brace Jovanovich, 1971), 168, see also 154–77. I am not seeking to make a statement about Sidney's philosophy. On this general critical trend, see Gerald Graff, *Poetic Statement and Critical Dogma* (Chicago: University of Chicago Press, 1980).

[2]Maurice Blanchot, *L'Espace littéraire* (Paris: Gallimard, 1968).

what I call "the central debate," which is the debate over the possibility of, and procedures behind, a valid interpretation of a literary work. Although the surfacing of this issue as a major and explicit theoretical problem is relatively recent, the same kind of issue has long been central to the most specific and particular interpretive endeavors. Moreover, the main contours of this debate are well known at this point, and there are few surprises to be expected in the literature. But this awareness has not prevented such discussions from going on, locked within what appears to be an inescapable impasse. Indeed, to acknowledge the necessity of failure would at this point seem to be the only way to get it right.

My own response is to try to cut deeper than this. Thus I ask what previous decisions, practices, and implicit assumptions have set this problematic up in a way that is inherently aporetic. The impasse, it seems to me, is wholly contingent upon these assumptions. Hence it is crucial to take up the issue of the ways the literary object and the knowledge we have of it have been defined. What is needed in this regard, what is in fact already well under way, is an understanding of the steps that were taken in the construction of the literary object as we know it—*and fail to know it*—in the present literary disciplines.[3] The conceptual requisites of a disciplinary model of knowledge are an important key to grasping the steps whereby a certain kind of object has been singled out from among a broad range of possibilities, at the same time as a certain kind of approach and attitude has been enthroned above others. In this light, it becomes crucial to pursue an analysis of the research practices that have been institutionalized in this form, with the goal of moving on to examine the emerging alternatives. In this regard I stress the central role played by the notions and impulses of a certain aestheticized interpretive practice in the way people conceive of literary knowledge. I try to demonstrate this point in an examination of some typical structures of literary explanations, structures isolated in relation to some discussions of explanation drawn from the philosophy of science. This analysis has the goal of showing the need for moving beyond this interpretive framework toward other perspectives. I lend my voice to the call for a sociological and historical framework for the study of literary phenomena, freed from the aesthetic a priori that guides and distorts so much of the interpretive endeavor, locking it

[3]See, for example, Gerald Graff, *Professing Literature: An Institutional History* (Chicago: University of Chicago Press, 1987); and Anne Mette Hjort, "The Interests of Critical Editorial Practice," *Poetics*, 15 (1986), 259–77.

into an impasse. *Which* sociological and historical approaches can best guide such an enterprise is another question, of course, and on that score I have only a few tentative suggestions.

Yet something remains missing in a sociological scheme of things. This kind of approach seeks to reinsert the literary personae, voices, and props within the actual social dramas where they emerged, and where they had their first consequences. Yet in so doing, it frequently has the effect of devalorizing its objects. Indeed, for some critics, to say that such an approach treats literature as an object is already to reveal what is wrong with it. Although this kind of complaint is often made for spiritualistic reasons, there is a grain of truth here, for how can a sociological reduction of the literary fact retain a sense of the value and importance of its object? More specifically, how can it sustain the idea of the cognitive worth of literature and its study? Literature, after all, would only be as important as the role these practices of writing and reading really exercized in history (and not in history as it is constructed as a series of "intellectual" or "cultural" artifacts and moments, which is yet another product of an idealist bias).

It is in response to this question that I will be stressing the value of a certain kind of interdisciplinary research, one where the critical analysis of literary works can have an invaluable heuristic function. More specifically, my claim is that the literary theorist and critic can make very real contributions to research programs within the social and human sciences by means of a strategic and selective analysis of literary documents, an analysis organized in function of explicit and well-delimited principles of pertinence. The aim of such analyses is to refine and complexify the models underwriting research programs within the human sciences; in such a context, the literary text can contribute to the crucial process of hypothesis formation. To make this path of research more tangible, one need only begin with a remark made by Castoriadis, who notes that contemporary economics is implicitly based upon a psychology of individual behavior that a cheap novelist of 1850 would have rejected as overly simplistic.[4] It seems to follow that the psychological insights in the great novels offer a set of initial assumptions and models far superior to those of orthodox economics. This is, I think, correct. Here, then, is an example of a topic upon which previously separate lines of research converge, a topic where the literary and the social scientific

[4] "Science moderne et interrogation philosophique," in *Les Carrefours du labyrinthe* (Paris: Seuil, 1978), 147–217, esp. 187.

meet. At the present time, there are real signs of progress toward a constructive interdisciplinary cross-fertilization in many areas of research.

One last preliminary remark. What follows is not an attempt to provide a comprehensive, normative discussion of our possible relations to literature. I believe that the question of literary-critical knowledge is to a significant extent distinct from the question of literary "experience" as a whole, although I would of course not want this distinction to be taken as an absolute disjunction of the two. The reader, then, is reminded that my focus is on literary critical writing viewed as *research.*

CHAPTER 6

Literary Explanations

In spite of the fact that current debates within literary theory have attained a vertiginous degree of verbal complexity, it is possible to characterize the key issues with a certain economy. In fact, one may venture that there is really only one fundamental issue, and that the others arise purely in relation to it. This key issue is the question of the validity of interpretation. Anything approaching an exhaustive discussion of the literature on this subject would require a voluminous study, for every serious critical theorist responds to this problem in one way or another. In this chapter, however, I do not attempt to engage in an extended dialogue with today's most prominent figures. I do not think that the best way to organize the analysis is to proceed from one major critical theorist to the next, conducting critical author studies on each. In that manner, critical theory only mirrors the most typical procedures of the kind of criticism it should be investigating. My goal, instead, is to present arguments pertinent to what I perceive as the central questions: What is the nature of literary knowledge? What are the predominant types of literary-critical knowledge today? To what extent are these types of literary criticism adequate? How do these types of knowledge stand up in relation to other models? What are the alternatives to the current patterns of inquiry in the literary disciplines? These are the crucial issues, and the literature of critical theory is of significance in relation to them, not for its own sake. For this reason, the literature is not foregrounded in what follows, which does *not* mean that my arguments bear no relation to it. Anyone remotely familiar with the debates can see that various names and titles could be provided as instances of the positions under discussion.

The Central Debate

What, then, is the question of the validity of interpretation about which the central debate spins? A simple scenario, which unfortunately is not pure fiction, sets the scene for my response:

A professor of literature gives a copy of a single poem to a group of people—students, educated professionals from different walks of life, other literary scholars, as well as colleagues from other disciplines. These people are asked to read the work, and then to write down what they think it means. The question may not be quite so broadly put, but the central thrust remains this: What is the message in the bottle?

Next the results are analyzed. The answers, it turns out, diverge wildly, and some flatly contradict each other. On this there is no serious dispute; clearly, one interpretation says the poem said *x*, another says it meant non-*x*, a third *y*, and so on. The next step, which seems to follow automatically, is to ask how these contradictory and divergent responses can possibly be answers to the question about the poem's meaning. Some of them have to be wrong, and others, right, it seems. How can we know?

The central debate concerns precisely this last question: How can we know which interpretation of a literary work is right? The many subordinate issues follow upon this one. They have to do, for example, with the nature and status of the norms of argument that should govern such an inquiry. They also concern the kinds of evidence available within such an inquiry—and most of all, the possibility of establishing criteria pertaining to the relative significance or pertinence of different kinds of evidence. Here one may plot the arguments over the possibility of defending some definition of the privileged context of interpretation, as well as the related quarrels over the status of the author's intention, the various attempts to detect typical fallacies, and general definitions of legitimate and illegitimate interpretive moves. Here as well we find the issue of authority—that of the artist, critic, or institution. Intimately connected to the same issues are all of the thorny problems having to do with the selection of criticism's units of analysis. Is it enough to explore the meaning of the poem alone? Should it be bound up within a corpus, genre, period, or other broader classificatory concept? Clearly there are a lot of extremely complex and far-reaching details to be dealt with in these subsidiary discussions.

In more recent trends, the central debate has given rise to strains of thought that present themselves as radical alternatives to the whole question. Drawn together, they can be seen to assert the impossibility of getting a definitive answer in the terms in which the question has been posed most frequently. To the question, How can we know which interpretation of a literary work is right?, one responds that we *cannot* know which interpretation is right. The discussion then focuses on what criticism should do in the wake of its renunciation of the quest for interpretive validity. Here there are subdivisions to be noted:

1. We cannot know which interpretation is right, and for demonstrable reasons: (a) because the work has in it essentially diverse and contradictory meanings, a kind of radical ambiguity; (b) because the only meanings the work has are those people read into it, and people read meanings into a work in function of their own frameworks—these frameworks being essentially different. The evidence is overdetermined by the framework, and reading is a kind of ventriloquism.
2. We cannot know which interpretation is right, because we cannot know how we could know. We cannot know how we could know because we cannot know . . . and so on. There is no valid metalanguage, so critical theory is ventriloquism too.

It may be remarked that, while the latter position closes down the inquiry (and thus typically is followed by the idea that critics should engage in something other than inquiry), the first two types of position seek to displace the interrogation and can do so in varying ways. In other words, to settle on point (2) usually means giving up on the knowledge claims of literary criticism, at which point its value, if it is still thought to have or to require any, must be stated in other terms; literary criticism is worthwhile, for example, because it subverts the philosophical desire for knowledge (theory as fun, politics, art). To settle on point (1), however, typically means that the orientation of critical knowledge is simply shifted. The *kind* of knowledge that it is expected to arrive at alters with this shift, and many different proposals have been offered. After taking one step back, this type of approach sets out to explain the conditions behind the diversity of readings.[1] Instead of trying to discover the true meaning of a

[1]Anything approaching a complete bibliography of the studies that belong within this trend would require a book-length effort. One should of course begin by citing

work—or even the broad range of its true meanings—the critic turns to the study of the determinants of its reception—be they historical, psychological, linguistic, political, or all of these.

Such, then, is a summary version of the state of the art, a somewhat caricatural version, but not so much so as to be irrelevant. Of course I must anticipate an objection—that of the critic who, on reading this, is quick to protest: "But wait a minute, this central debate of yours—it isn't *my* problem at all, and it's not being conducted in any of the journals *I* take seriously. Rather, it's the preoccupation of the *theorist*." Yes, it is true that there are many literary scholars whose work is not oriented by the assumptions that underlie the central debate. But again, there are many whose work *is* directly related to the issues raised in this debate, that is, those engaging in stating what literary texts mean, and who might be curious as to the question of the validity of their results. We are probably right in saying any number of bad things about the central debate, but not that it is only the product of some critics' penchant for theorizing. Another way to put this is to say that the authority of critical interpretations is indubitably an effective fact within the institutions where literary critical knowledge is created and taught; the object of the central debate, then, has to do with the legitimacy of this authority, the grounds of its claim to validity. In other words, if you are a professor of literature, the status of your professional activities is directly related to the issues considered in the central debate.

What I want to say to the reader—as emphatically as possible—about this central debate is simply this: it is a mistake to think that one must choose from among the positions that it typically embraces. The central debate, as well as all of the stances within it (including its skeptical implosion), is a vast attempt to respond to a question that is poorly posed and badly misunderstood. It is this pseudoquestion that is the conceptual source of what is essentially an impasse. It is a mistake to choose from among these positions, then, because they each try to give an answer to what is really a kind of epistemic trap. If I *had* to give my opinion on the "question" as it appears in the central debate, I should have to say that only the purest skepticism (the second response above) would be warranted, for the question is set up in a way that makes any determinate inquiry and

---

Roland Barthes, *Critique et vérité* (Paris: Seuil, 1966); one other clear and concise example is Mircea Marghescou, *Le Concept de littérarité: Essai sur les possibilités théoriques d'une science de la littérature* (The Hague: Mouton, 1974).

answer impossible. But when skepticism starts to look as if it were warranted, we have reason to think that there is something wrong with the basic orientation of the inquiry. To give a precise analysis of what is wrong is no simple matter, and I shall be attempting to do so at some length in a moment. But in order to set the course of my argument, I shall first compare the little reading experiment described above with another scenario.

A professor of physics walks into a classroom filled with people. Some of them are colleagues and students of physics, others are chemists, engineers, philosophers, literary critics, doctors, and lawyers. The professor brings along a metal stand, which supports a fixed peg, on which there hangs a pendulum. Pointing in the direction of the device, the professor says: "Now explain this system. Conduct whatever tests, observations, or measurements you like, but explain this system." (In a crueler version, the physicist orders the class to tell what the system means.)

Later the professor examines the results. The responses written by nonscience students, or by those who have forgotten what they once learned, present an amusing variety of approaches. Some do not focus at all on the kinds of questions a physicist might ask about a physical system. They are, as the saying goes, impressionistic. In a process that is curiously referred to as that of "free association," the swinging of the pendulum is linked to the idea of a clock, which gives rise to reveries about time, mortality, the transience of all things, the fiction of Edgar Allan Poe, the Spanish Inquisition. Others offer a kind of commonsense approximation of what they take to be a scientific mode of explanation; one genius has scrawled: "What goes up, must come down. This pendulum is a what, and went up. Therefore, it went down." But it is the people who are well versed in science who are most baffled by the problem. Which system? they wonder. Some choose to narrow this vague business down by hearkening back to some example of a pendulum in an elementary textbook, and in this way they are able to provide a banal but seemingly acceptable answer. Their basic training makes it possible to slip in a more precise question, one providing a definition of which variables are to constitute the system under consideration. Typically, they hearken back to the highly idealized model of a simple pendulum, described as a point mass on the end of a weightless rod of length *L*—thereby excluding from consideration a multitude of the factors pertinent to an analysis of the physical or compound pendulum. In terms of the selected variables, the physics students can specify something that might need explaining. Some, for example, simply regurgitate the equation for the calculation of the period of the pendulum: $T = 2\sqrt{l/g}$ (two times the square root of the quotient obtained by dividing the length by the gravitational constant). But those who take the professor's question as a real challenge to do some science are at a loss, for it was wholly unclear which variables were meant to be explored. The professor, after all, only pointed in the direction of the pendulum, and did not say anything that could narrow the level of analysis, thus leaving the students to ponder the many possibilities. Specific gravity? The film of moisture on its surface? Electrical conductivity? Mass? Temperature? Chemical composition and impurities? Tensile strength? If the trained scientists were given the time really to get to work, their analyses would have presented a bewildering diversity and complexity. Among the reports turned in is that of a biologist, who identifies some bacteria found on the device (but who is at a loss as to what he is supposed to find out about them).[2]

The comparison between these two fictive examples is introduced here to emphasize both the similarities and the differences between the physical and literary cases. A first and crucial point that the analogy evokes is this: if, in the case of the little "reading experiment," the results amount to a chaotic and bewildering diversity, this does not warrant us to conclude that a more exact interrogation of a text cannot yield far more exact results than those arrived at in the example (such conclusions are frequently drawn in literary theory). In this regard, the cases are not essentially different, for a similarly slack manner of posing questions in physics gives rise to a highly chaotic group of responses. The responses to the physicist's vague question include some bits of knowledge only insofar as the physics students ignore the question's vagueness. Taking their cues from the context, they make the question more precise by adopting the priorities of some canonized model of physical theory. It is crucial to note as well just how idealized and selective this theoretical model is; the actual shape and composition of the rod is replaced in the model by the fiction of a weightless line, the weight at the end becomes a mass point, and the students approach the details of the real pendulum only in function of the model of the simple pendulum. Only a theory, the word being taken here in a very broad

[2]This example is discussed by William Ross Ashby in *An Introduction to Cybernetics* (London: Chapman & Hall, 1956): "Any suggestion that we should study 'all' the facts is unrealistic and actually is never made. What is necessary is that we should pick out and study the facts that are relevant to some main interest that is already given" (40).

sense, specifies a question that can be given anything like a precise result.

In this regard, the key point can be underscored by means of another analogy. Let us imagine that someone walks into a cartographer's shop and says, "I want the map of Denmark." The mapmaker replies, "What sort of map would you like?" But the customer insists: "The map of Denmark." The cartographer might now assume that the person making this unusual inquiry is only a motorist who, being in a hurry and preoccupied with his own affairs, wrongly takes for granted the idea that the roadmap he requires is the canonized model of what a map should be like. The fellow ignores the fact that in the cartographer's drawers are maps designed for wholly different interests, maps, for example, that cater to the interests of a fisherman. In this case, the misunderstanding is resolved when the mapmaker quickly provides the standard highway map and the customer goes on his way. But imagine that the customer insists: "No, this isn't *the* map of Denmark. I've seen ones like this before; there's a whole lot of stuff missing." The cartographer, who is amazingly patient, sees a chance to get a clue as to what the fellow is on about, and asks a leading question: "What kind of missing things were you interested in, sir?" After a pause, the fellow answers, "All kinds." At this point, the cartographer is helpless. Had the fellow been able to narrow down for him a kind of orientation toward the domain of Denmark, a set of criteria or topics specifying a principle of pertinence or selection, he could have chosen or set out to make a map for this gentleman, beautiful maps, maps of extraordinary accuracy and detail. The complete map for the mushroom hunter. A detailed map of restaurants and inns. An exhaustive map of golf courses. A host of nautical and road maps. Maps on fish and fowl, castles and museums, parks and industries. But a request for *the* map puts a halt to this science, and the cartographer is speechless.

The point of this little parable is that the man's request is inimical, not only to the science of cartography, but to science as such. In Marx Wartofsky's view, science is an oriented map-making activity.[3] My expression is perhaps redundant, for to say "map" is to imply the necessity of a selective and oriented reference to a domain, the sum totality of which cannot possibly be re-presented. To think that such

[3]Marx W. Wartofsky, *Conceptual Foundations of Scientific Thought* (New York: Macmillan, 1968), 123–37. One may also usefully refer to Leszek Nowak, *The Structure of Idealization: Towards a Systematic Interpretation of the Marxian Idea of Science* (Dordrecht: Reidel, 1980), and particularly to his distinction between idealization and the Weberian notion of ideal types, 41–53.

a total re-presentation is the goal of any science is what Richard Rudner calls the "reproductive fallacy," and he mocks this idea by associating it with the notion that it is the goal of science to give the "taste of the soup."[4] Science does not strive to re-produce the taste of the soup; nor does it seek to re-produce the totality of a domain's determinations, seeking to obtain the pure presence of the object. Here we need to recall Alfred Korzybski's maxim, which Gregory Bateson was fond of citing: "The map is not the territory."[5] But to say this does not open the door to any of the absurd skepticisms whereby no map can be rightly said to refer to anything at all; some maps, particularly those that are carefully and explicitly oriented and checked against the territory, do refer to real features of the domain; this reference, however approximative and selective, is the source of truth and falsehood. (Thus a map that shows a bridge connecting Fanö and Esbjerg is, until further notice, an incorrect and false representation of the domain.) The "maps" provided by science, then, are necessarily selective and abstractive, but these features do not subvert the possibility of reference; what they subvert is only the myth of a total and full re-presentation of the real (in its "rigorous purity" and "absolute singularity," for example). As Wartofsky points out, a total reference system would involve a proper name for every item in the territory—a nominalist fantasy experienced by Borges's Funes. The reference systems of science are, on the contrary, a matter of what Wartofsky calls a "selective confrontation" of properties, some of which are chosen while others are ignored. Observations of sameness and difference, then, cannot be absolute—even if they can be true. One thing or system is referred to as "the same" or as "different" *in some respect*, or in regard to *a certain property*. An *x*, which is already grasped not as a unique particular but as an instance of a type, is noted as being more like a *y* than a *w* is like a *z*. In other words, only certain properties, and the relations between them, are singled out as pertinent in any given mapmaking endeavor, and such an endeavor is organized in function of its theoretical selectivity and abstraction, for these conceptual moves alone can provide the necessary orientation. As Wartofsky puts it, "everything is like everything else in an infinite number of respects and different from everything else in an infinite number of respects."[6]

[4]Richard S. Rudner, *The Philosophy of Social Science* (Englewood Cliffs, N.J.: Prentice-Hall, 1966), 69–70.

[5]Alfred Korzybski, *Science and Sanity: An Introduction to Non-Aristotelean Systems and General Semantics* (Lakeville, Conn.: International Non-Aristotelean Library, 1933), 58.

[6]Wartofsky, *Conceptual Foundations*, 129.

As was suggested above, it may be redundant to speak of science as oriented mapmaking insofar as all maps are necessarily oriented—if by this is meant only "selective" and "abstractive." No map, phrase, utterance, depiction, message, or semiotic system can possibly *be* the territory to which it refers. Yet the expression is no longer redundant if we understand "oriented" to mean "deliberately and systematically delimited and organized," in which case the phrase could serve to help distinguish between scientific and other types of maps. Nonscience, then, would *in part* be a matter, not only of those maps that refer to nothing at all, or that refer wrongly, but also of those that refer in an unsystematic, broad, indefinite, and nonmethodical fashion. The scientific map would be one guided by an explicit theoretical knowledge of the principles behind the construction of a model, this model or language being organized by means of selective inclusions and exclusions of what it is deemed to be pertinent to depict and to explain; it would also be a map that had been adequately checked against the domain. If the physics students were not thrown off course by the sight of a pendulum, it is because they have already been shown a map upon which this kind of device serves as an example for a particular kind of motion. The theory is already in place, has frequently been employed, and generally makes possible reliable results whereby particular variables selected from the domain are plotted in terms of a general pattern or law.

But is there no theory, then, behind the literary professor's question about the meaning of the poem? This is, I think, the crucial question today for an analysis of the nature of literary knowledge. The answer is far from simple. If behind every mapmaking activity there is at the very least an implicit orientation, the responses qiven to the question of the poem's meaning are clearly not purely random, a sheer chaos lacking any organization whatsover. There would, then, be something redundant about the ways the question was understood and answered, a pattern subtending the diverse and contradictory results one may observe at another level. This is, at the very least, a notion that should be explored. At the same time, it is clear that the reading experiment as described is an example of a severely disoriented form of mapmaking if compared to the more successful instances of scientific knowledge. The members of the group are assigned a vague task, and the vagueness and diversity of the results are no surprise in the light of this first step.[7] The guiding

[7]A similar comment is made about this kind of reading experiment by Max Black, "The Radical Ambiguity of a Poem," *Synthese*, 59 (1984), 89–107. Black notes that

assumptions are only implicit, left to be filled in by the individual response. The cartographer is shown a poem and asked to make a map of its meaning; is it a wonder that what his sketch shows us is most of all what is on his own mind?

I have just suggested that in order to come to terms with the problem of the central debate in critical theory it is necessary to explore the orientations that underwrite a range of interpretive activities. At one level of analysis, it appears that these interpretive activities have nothing to do with scientific knowledge at all; the inconclusive and fragmentary nature of the results, for example, could be held to lend ample support to such an observation, and one might go on from there to conclude that the mapmaking in question is totally chaotic and disoriented. This is the level of analysis at which it makes sense to speak of a crisis in the discipline, particularly if one has in mind something similar to the following notion of what it means for a field to be in crisis: "A research field is said to be in a state of *crisis* if it is stagnant, or is dominated by a single narrow school, or is fragmented into several warrying [sic] schools, or is split into many narrow and weakly related specialities, or some of its own accomplishments are threatening its dominant conceptual framework."[8] Literary criticism today is certainly not stagnant, but anyone who thinks its schools are not "warrying" has not taken a look at the literature.

At the same time, is it the case that critical work is so very disorganized? At another level of analysis, it may be possible to perceive that critical mapmaking is highly structured and can be characterized in terms of certain invariant attitudes and assumptions. But if this is so, what are these assumptions? Can they be made explicit? How is it that a common set of assumptions gives rise to such diverse and fragmented results? Are these the best assumptions that could guide literary critical research? In what follows I grapple with these questions. In the present chapter I focus primarily on the predominant interpretive assumptions, and in the next I take up the alternatives to them.

In a sense, part of my endeavor reveals an affinity to one of the broad kinds of positions mentioned above. With the semiologists and others who step back from the practice of literary interpretations in

---

readers' inability to clarify ambiguity is related to a "lack of clarity about the point of reading fiction or hearing a play" (92); Black's quasi-Wittgensteinian assumptions, however, lead him to reach very different conclusions from my own.

[8]Mario Bunge, *Treatise on Basic Philosophy*, vol. 6: *Epistemology & Methodology II: Understanding the World* (Dordrecht: Reidel, 1983), 180.

order to explore the conditions of possibility of literary meaning, I agree that some of the goals of critical theory make it distinct from the critical interpretation of works. An epistemologically oriented critical theory should not give itself the goal of finding an algorithm for generating correct interpretations of literary works; even less does it stage a courtroom scene in which it will sit and judge individual literary experiences. In another sense, however, my undertaking here is quite different from the theories that try to build models of literary competence, or which elucidate in a general manner the processes of literary cognition. The first major difference has to do with my emphasis on literary *knowledge*, an emphasis that differs even from that of many of the literary theorists who still believe that such a notion can be defended. I would stress, for example, that the possibility of knowledge *in* literature should not be limited to such broad notions as cognition, meaning making, aesthetic experiences, or *Bildung*. That meaning is possible does not mean that it is true, valid, or important. Nor is a "meaningful" belief equal to "knowledge," except in a loose and self-serving acceptation of the latter. In regard to the possibility of a critical knowledge *of* literature, I contend that the knowledge in question should not be limited solely to the construction of grammars. It does not strike me as a particularly good idea to try to write a rigorous grammar that models the present practice of aesthetic hermeneutics; not only is it immensely difficult to write a real (algorithmic) grammar of anything but the simplest artificial language (and highly abstractive semantic representations of natural ones), but it is not clear why one would prize such a grammar. My own view is that theorists should be interested in the question of the status of critical research in relation to the predominant model of knowledge, which is that of scientific explanations in the natural and social sciences.[9] I do not think that this question is

[9]A well-informed and far-reaching work that cannot go unmentioned in this context is Charles Altieri, *Act and Quality: A Theory of Literary Meaning and Humanistic Understanding* (Brighton, Sussex: Harvester, 1981). Although I applaud Altieri's remark to the effect that the question of literature as a form of knowledge is fundamental (270), I believe that there are more alternatives than his basic triad of skepticism, scientism, and his own philosophical reconstruction of what he at one point speaks of as the "humanistic pieties." Altieri contends this: "Unless we can recover the *force* that writers intend by their efforts as artists and interpreters of action, I do not think we can justify appeals we make to their authority and wisdom. And without such appeals, I doubt the humanistic education or its objects have much claim upon the attention of society" (1). This view makes some sense to me, yet I wonder if the real thrust of the question of literary knowledge has anything to do with appeals to authority at all; I also wonder whether the appeals in question are really justified by the conclusions Altieri is able to reach by turning to Nelson Goodman's aesthetics. Altieri is right to reject any *simple* propositional and empiricist approaches to the

adequately addressed within the confines of the present disciplinary debate over the validity of interpretations of literary works. The assumptions and orientations that have guided that debate, assumptions having to do with the very nature of validity, for example, are simply inadequate to that task. Semiotic and other theories which give themselves something called "literary competence," and with it, a literary understanding of what validity and knowledge are all about, do not take the necessary distance from the assumptions that have underwritten the central debate. But how can such an analysis proceed? I begin with a few remarks of a general nature, as I think it crucial to have a broad perspective on this issue.

Idioi topoi, koinoi topoi

What is a science? What is the relationship between the idea of science and the idea that real knowledge is divided into separate disciplines? It seems to go without saying that rigorous knowledge is that which has been created through an increasing specialization and differentiation of both the objects and the methods of inquiry, or, in other words, that genuine knowledge—precise and well-oriented knowledge—is disciplinary knowledge. Yet what is a discipline? Some elementary points related to these issues are made by Aristotle when, in his *Rhetoric*, he distinguishes between *idioi* and *koinoi topoi*. The former are argumentative devices belonging uniquely to a particular, well-defined class of things; the latter are general topics having no special subject matter, commonplaces that "apply equally to questions of law, the natural sciences, politics, and many other

---

knowledge *in* literature, but I fear that his otherwise excellent arguments for the "cognitive" benefits of reading good fiction still fall short of the stronger claims that are really needed to justify more specific knowledge claims in regard to literary-critical research—and not their general authority. He notes, for example, that "literary authority derives from making problems believable, not from solving them" (237). This and other arguments like it in Altieri's book strike me as being appropriate as an extremely broad justification of the benefits of literary pedagogy, but it is unclear how they support the validity of any specific critical inquiries. Even on the former issue one may have doubts about such attempts to legitimate literary knowledge, particularly within a North American context, where one may be skeptical concerning the literary culture's putative role in promoting "sensibility," "taste," and "tact," the ability to "register subtle features" and "understand a range of attitudes." In any case, one may well believe that these terms refer to real virtues, and even that reading literature may help to foster them; it is, however, unclear how this kind of cultivation of personalities provides a solid basis for speaking of critical *research* in the literary disciplines. Is this not like trying to justify the global validity of dynamic psychological theorizing by referring to therapeutic consequences these theories could, in principle, bring about?

disparate things."[10]. Aristotle says that for a discourse or *logos* to be a science, it must construct its arguments primarily with the former type of *topos*; that is, it must possess a well-delimited object domain, one it shares with no other discourse, and it must address this domain with the arguments and terms belonging specifically to it. In this manner, both the knowledge and its objects are determinate (*peras*), that is, well bounded and having clear and distinct identities not confused with anything else. We may note that discursive principles identified in this manner are a matter of both the objects being discussed and the terms and arguments used to refer to them; just as an essence or identity is distinguished from what it is not, so is an autonomous discourse dealing with this class of objects constituted in the same gesture.[11] To the specific, determined object corresponds the specific, disciplinary discourse that masters and knows it. A whole series of disjunctions follows from this initial gesture: rhetoric, an art or *techne*, is distinguished from the class of autonomous sciences, for it is perceived as a separate collection of sayings and devices that can be applied by anyone in a discussion of anything—which is not, for Aristotle, to say that rhetoric is necessarily sophistical, or that in its essence it betrays the superiority of the true and good arguments.[12] Science, the domain of exact entities and relations, and of the exact language used to refer to them, is separated from the *public* sphere of discourse. In a rigorous science, arguments and knowledge proper to the subject matter result in conclusive, binding demonstrations, for anyone who moves through the orderly steps of the argument must agree with the result. In rhetoric, on the contrary, the *logos* is influenced by the conditions of persuasion (*pathos* and *ethos*), which are external to the proper subject matter of the discussion. Rhetoric, then, is the realm of opinion that may or may not be well founded, whereas *episteme* is a matter of domains of binding truths.

Although Aristotle distinguishes between rhetoric and science in this manner, we must remember that in his perspective the commonplaces of rhetorical persuasion are not necessarily inimical to

[10]Aristotle, *Rhetoric* (Cambridge, Mass.: Harvard University Press, 1926), 1358a.21.

[11]Aristotle gives no explicit definition of what is meant by a *topos*, and commentators diverge on whether it should be taken primarily as a rule or law of argumentation, principle, guiding proposition, premise, or other. See W. A. de Pater, "La Fonction du lieu et de l'instrument dans les *Topiques*," in *Aristotle on Dialectic: The Topics*, ed. G. E. L. Owen (Oxford: Clarendon, 1968), 164–88.

[12]This crucial point is argued by William Grimaldi, *Rhetoric I: A Commentary* (New York: Fordham University Press, 1980), and *Studies in the Philosophy of Aristotle's Rhetoric* (Wiesbaden: Franz Steiner, 1972).

the task of arriving at correct and fair judgments, and that rhetoric remains an essential and unavoidable form of deliberation that could never be replaced by a science of the political. In this sense, the "common" discourse also has its own "proper" and specific domain, the public sphere where matters of a general interest come into dispute. We should also remember that in Aristotle's view it is the most general discourse of all, first philosophy, that ultimately gives the special sciences their basis.[13] Aristotle remained supremely confident about the unity of knowledge, and of the cosmos it was to describe—and he was, after all, the world's greatest interdisciplinary thinker. At the same time, we cannot fail to recognize that the gesture of disciplinarity is sketched by him with a remarkable clarity in the passages from the *Rhetoric*; moreover, Aristotle's emphasis on the internal relation between science and disciplinarity was certainly not lost on the tradition that followed in his wake. Kant, for example, restates very clearly the assumption that knowledge is scientific to the extent that it is disciplinary, that is, divided into clear and distinct discourses and objects: "If one wants to present a knowledge as a *science*, it is first of all necessary to determine exactly its distinctive characteristic, what it has in common with no other knowledge, what is particular to it; failing this, the boundaries of all the sciences run into each other and none of them can be treated fundamentally and in conformity to its nature."[14] It would be easy to multiply such citations, and in a moment I indeed document the extent to which a similar notion has guided many efforts to constitute a literary knowledge following a disciplinary model. One note of caution, however, is in order: the assertion that knowledge is scientific to the extent that it is disciplinary is counterbalanced, both in Aristotle and in Kant, by an emphasis on the necessity of connecting the special sciences to the larger framework of philosophical reflection, a framework where similarities as well as differences are perceived. It is particularly difficult to see how a singleminded stress on the specificity of a science's objects and methods can be reconciled with the crucial role played by formal languages within natural science, for in their very formality these conceptual tools are anything but specific to any ontic region. Perhaps it would be prudent to stress that what we are presenting here is only one part of what makes a discourse scientific, a part that is highly pertinent—perhaps

[13]Giovanni Reale, *The Concept of First Philosophy and the Unity of the Metaphysics of Aristotle*, ed. and trans. John R. Catan (Albany, N.Y.: S.U.N.Y. Press, 1980).
[14]*Prolegomena to Any Future Metaphysics*, trans. L. W. Beck (New York: Bobbs-Merrill, 1950), 13.

even necessary—without being exhaustive or sufficient. Moreover, to anticipate some of my conclusions, I think it worthwhile to ask whether the literary discipline's stress on the specificity of its object has not in certain applications of this notion led to a distortion of the very motives behind such a gesture—a distortion, that is, of a type of inquiry within which a search for specificities can make sense. Moreover, to search for *the* specificity of literature may be like asking for *the* map of a domain that has no separate existence apart from the human beings who think of it.

It is clear, at the very least, that assumptions about the relation between knowledge and disciplinarity have been instrumental in critical reflection over the status of literary knowledge. Some examples are in order. A characteristic statement is Roman Jakobson's famous formulation of the basis of literary research, first uttered in 1919: "The object of the science of literature is not literature but literariness, that is, what makes a given work a literary work."[15] And here is how Wellek, some forty-five years later, put it: "If we want to arrive at a coherent theory of literature, we must do what all other sciences do: isolate our object, establish our subject-matter, distinguish the study of literature from other neighboring disciplines."[16] And it has been added more recently that "the basic problem of literary study, the one to which all others in the field can be traced, is the lack of a precise conception of the thing being studied. Indeed, one could argue that there can really be no field at all until there is an agreed-upon definition of literature."[17]

What is at stake in these calls for a definition of the literary object?[18] There seem to be at least two major facets to the issue. On the one hand, there is the problem of stating what kind of item is to be included within the range of "the literary" and of then giving a procedure for determining which "objects" are to be included and

[15]"L'objet de la science de la littérature n'est pas la littérature mais la littérarité, c'est-à-dire ce qui fait d'une oeuvre donnée une oeuvre littéraire"; Roman Jakobson, *Questions de poétique* (Paris: Seuil, 1973), 15.

[16]Cited by Bennison Gray in *The Phenomenon of Literature* (The Hague: Mouton, 1975), ix.

[17]Gray, *Phenomenon of Literature*, 1.

[18]A concise survey of some of the main texts is provided in Thomas Aron, *Littérature et littérarité: Un essai de mise au point* (Paris: Belles Lettres, 1984). Interesting historical surveys of the term's semantics are those of René Wellek, "Literature and Its Cognates," *Dictionary of the History of Ideas*, ed. Norbert Wiener (New York: Scribner's, 1973), 3:81–89; and Roger Escarpit, "La Définition du terme 'littérature,'" *Proceedings of the IIIrd Congress of the International Comparative Literature Association* (The Hague: Mouton, 1962), 77–89. See also Tzvetan Todorov, "The Notion of Literature," *New Literary History*, 5 (1973), 5–16.

which to be excluded. Second, there is the problem of saying something fairly precise about what the science of literature wants to know about these objects, for surely one does not want to propose that the literary discipline undertake an exhaustive explanation of things literary; such an attitude takes us back to the "pendulum as such." This latter point may be driven home by means of an example. Let us suppose that, following some response to the first problem mentioned above, it has been decided that *Moby Dick* belongs squarely within the class of literary works. Let us also imagine that some horrible positivists rig up a machine that can do calculations of the frequencies of words in this text, and that it is possible to relate these to well-established data about average word frequencies in the written documents of the period, so that it is possible to make some fairly rigorous statements about the particularities of the work's lexical selection.[19] Is this a *literary* knowledge of *Moby Dick*? My point here is not to answer that question, but to suggest what other kinds of questions must be answered before it can be sensibly dealt with. We clearly have in this example at the very least a rough sketch of a bit of knowledge that has been chipped off of the monumental *Moby Dick*, and this knowledge seems to be pretty indubitable in itself. But is there any such thing as knowledge in itself? To give a real response to the question of the status of this kind of research, we need a clearer sense of what broader and deeper interests and questions guide it. Literary critics would most frequently consider such research to be, at best, an auxiliary tool. They would think that, in spite of their positivity and rigor, the results could hardly be thought to address the real questions one should pose about a literary masterpiece. But what definition of the real questions supports this judgment—a judgment with which I, for one, would be quick to agree? Following what implicit assumptions does a critic believe that literary knowledge cannot possibly be equated with any and all knowledge pertaining to a literary work? And please note that this assumption comes into play only after it has been assumed that it is possible to say what is, and is not, a literary work—no small task in itself.

The example of the analysis of *Moby Dick* sheds some light on the motivations behind the search for literariness and suggests why this search cannot reasonably be said to be the affair of only one branch of literary criticism, that of the formalists, Russian or other. The guiding questions are those of the literary discipline as such, for without a

[19]The example is anything but farfetched; see Robert Oakman, *Computer Methods for Literary Research* (Atlanta: University of Georgia Press, 1984).

response, without at least an implicit assertion that such a response exists (if only in the form of the mute and stern authority of the institution), there can be no coherent notion of literary research. What, then, is literature, and how, within it, is the specifically literary defined? The question is rhetorical. It is not stated, here at least, as a question to be answered—but as a question having assumptions, interests, and a history which must be explored. What seems clear to me is that we do not have an agreed-upon answer to this question, which is not to say that the institutions of literary criticism in North America and elsewhere are not heavily invested in a certain set of assumptions about the answer—assumptions that, like the institution of the teaching and study of modern literatures, have a very short history. In the remainder of this chapter I try to get at some of those assumptions, first through an examination of some theoretical statements made by critics, and second in an analysis of examples of concrete literary research.

Pure Aesthetics, or the Message and the Bottle

A useful point of entry for an investigation into the assumptions behind many literary interpretations is provided by the broad remarks made by Wellek in his appropriation of ideas taken from Ingarden, the formalists, and the traditions to which they in turn are indebted. Again and again, critical theorists have sought to isolate the properly *literary* attitude toward a work, an attitude that is consistently based upon some notion of an aesthetic variety of language defined in opposition to its scientific and everyday functions and modes of organization. A series of oppositions is drawn: literary language is connotative, figurative, while scientific language is denotative, literal; in literary language, the sign itself is foregrounded, in science it is not; literary language refers to a world of fiction and imagination, whereas the scientific usage does not; literary discourse is individual, personal; scientific language is impersonal. At the same time, however, it is necessary to defend the boundary of the literary on the other front, for if it must be disjoined from the utterances of science, it must also be separated from those of everyday life and of the sacred. This demarcational task is complicated by the fact that some of the features indicated as literature's difference from a crystalline science are obviously present in ordinary speech as well. Linguistic norms are violated in common conversation, where there are innovations, figures, puns, emotional and personal expressions, and creative moments. Is the difference, then, only a matter of degree? The notion of a purely aesthetic function whereby a complex verbal artifact is appreciated as an end in itself buttresses this boundary, for the quasi-literary figures of everyday life rarely have the same status. Nor in everyday examples of "display" or storytelling is there such an investment in "fictionality," in a feigned-but-suspended reference, as there is in a literary work of art.[20] But here I think it best to give a somewhat lengthy citation of Wellek's text, which is invaluable in its crystallized statements of assumptions shared by numerous critics, including those who may have long forgotten, or never read, his book:

> All these distinctions between literature and non-literature which we have discussed—organization, personal expression, realization and exploitation of the medium, lack of practical purpose, and, of course, fictionality—are restatements, within a framework of semantic analysis, of age-old aesthetic terms such as "unity in variety," "disinterested contemplation," "aesthetic distance," "framing," and "invention," "imagination," "creation." Each of them describes one aspect of the literary work, one characteristic feature of its semantic directions. None is itself satisfactory. At least one result should emerge: a literary work of art is not a simple object but rather a highly complex organization of a stratified character with multiple meanings and relationships. The usual terminology, which speaks of an "organism," is somewhat misleading, since it stresses only one aspect, that of "unity in variety," and leads to biological parallels not always relevant. Furthermore, the "identity of content and form" in literature, though the phrase draws attention to the close interrelationships within the work of art, is misleading in being overfacile. It encourages the illusion that the analysis of any element of an artefact, whether of content or of technique, must be equally useful, and thus absolves us from the obligation to see the work in its totality.[21]

A careful analysis of the implications of this passage could be a first step toward an understanding of the way certain aesthetic assumptions have oriented literary knowledge as well as literary ignorance. The next step in this understanding would involve a challenging of these assumptions, one meant to determine whether they are viable any longer, whether they can rightly be thought true, legitimate, useful, or of value. Wellek's passing comment that the terms

[20]Mary Louise Pratt, *Toward a Speech Act Theory of Literary Discourse* (Bloomington: Indiana University Press, 1977).

[21]René Wellek and Austin Warren, *Theory of Literature* (New York: Harcourt, Brace & World, 1956), 15–28, citation: 27–28.

he is amassing are age-old provides a starting point, for it seems crucial to situate the synthesis he proposes here in relation to the kind of magnificent historical overview for which he is so rightly credited. And in these terms, the age-old notions no longer seem quite so ancient, particularly when given the semantic twist that Wellek provides here. None is satisfactory on its own, Wellek rightly states, for the guardian aesthetics of the literary discipline is indeed a composite form, difficult to pin down. None, it is worthwhile to add, becomes a dominant literary-critical concept any earlier than the late eighteenth-century turning point that Wellek himself has ably pointed to as the moment of the emergence of a new definition of literature.

What, then, are the varied assumptions about the nature of the work of art that begin to emerge toward the end of the eighteenth century and became predominant not long thereafter? The terms Wellek foregrounds evoke an artist who is imaginative, creative, innovative, and in a profound sense antimimetic. His critic is not primarily interested in truth, morality, or any of the other "extrinsic" issues. The formalist's interests lie elsewhere. But where? Here I think it crucial to take up an argument made by Tzvetan Todorov, who defends the idea that the conceptual foundations of formalist aesthetics in general are to be traced back directly to the Romanticist inspiration.[22] I shall briefly restate his thesis and then provide a few comments of my own.

Todorov does not hesitate to provide a broad characterization of what he unflinchingly calls a "romantic aesthetics." He sees this doctrine as having been brilliantly prefigured in the little-known essays of Karl Philipp Moritz (but he also adds that other choices could have equally well been made). He thinks it possible to give a schematic characterization of the movement's main tenets and assumptions by means of a reading of the motifs crystallized by Moritz in a brief essay of 1785, "Versuch einer Vereinigung aller schönen Künste und Wissenschaften unter dem Begriff des in sich selbst Vollendeten."[23] How, then, does Todorov characterize the

[22]Tzvetan Todorov, *Théories du symbole* (Paris: Seuil, 1977); Todorov's historical account could be usefully supplemented by that of Paul Oskar Kristeller in his "The Modern System of the Arts: A Study in the History of Aesthetics," *Journal of the History of Ideas*, 12 (1951), 496–527, and 13 (1952), 17–46. Kristeller stresses the preromantic origins of the modern system of the arts, but this emphasis does not contradict Todorov's insight concerning the romantic bases of a certain doctrine of the autonomy of aesthetic structures.

[23]Although I am greatly indebted to Todorov's discussion, I permit myself a few chicanes; Todorov seems to think it highly controversial to cast Moritz in the role of

essential assumptions of the romantic turn in esthetics? He emphasizes, first, a major shift in the understanding of the work of art entailed by thinking of the artwork as poiesis rather than mimesis, the one a matter of creation, the other a matter of imitation. Second, in the romantic concept, art is "intransitive," it has no goal or function outside of itself; it has been "purified" and does not seek to instruct, to please, or to edify. Third, the romantic work is characterized by its "organicity," that is, by its strong internal coherence. These very features lead to a fourth, namely, that the work of art cannot be paraphrased adequately or exhaustively; its significance exists on its own terms only and as such cannot be translated into any other form or language. Stated positively, this impossibility of translation amounts to an infinite richness in that the interpreter can never exhaust the work, which thereby maintains a transcendent plenitude of meaning. Finally, Todorov isolates a fifth major tenet in what he refers to as an aesthetic *synthétisme*; in its holistic perfection, the work of art resolves the dualisms and dichotomies that scar the extraartistic domain. It unites matter and spirit, form and content, freedom and necessity, consciousness and unconsciousness, and so on.

Is Todorov's list well founded? In the case of Moritz, the answer is affirmative.[24] Moritz explicitly and forcefully rejects the inherited idea that the purpose of art is imitation and pleasure and points out that these notions do not distinguish artworks from other things that are merely useful. A tool, for example, has its end outside itself—in the user. A tool can therefore only achieve perfection through the mediation of something else. In art, the opposite holds. Moritz writes: "I observe it as something that is perfect, not in me, but in itself; thus as something that constitutes a whole in itself, as something that affords me a pleasure for its own sake."[25] In relation to art,

---

the announcer of the romantic and lashes out at the tendency to cast this innovative figure in the shadows of Goethe (which is where he tended to see himself—Moritz is said to have referred to the master as "God"). Although it may be true that Moritz has remained a relatively marginal figure, scholars have hardly been blind to the notion that certain of his ideas were to receive a fuller expression in Goethe, Novalis, Jean Paul, and August Wilhelm Schlegel. Joseph Nadler, for example, characterizes Moritz as "the first who can rightfully be called a Romantic"; cited by Mark Boulby in his *Karl Philipp Moritz: At the Fringe of Genius* (Toronto: University of Toronto Press, 1979), 192.

[24]Karl-Philipp Moritz, *Schriften zur Ästhetik und Poetik*, ed. Hans Joachim Schrimpf (Tübingen: Max Niemyer, 1962); In my discussion of Moritz, I follow Todorov in drawing heavily upon the 1785 "Versuch," but I also refer to two later essays, "Über Zusammenhang, Zeugung und Organisation" (1788), and "Die Signatur des Schönen. In wie fern Kunstwerke beschrieben werden können?" (1793).

[25]Moritz, *Schriften*, 3.

then, one shifts the goal away from oneself and places it back in the object itself. Moritz calls this an attitude of love, and adds that, in positing the absolute self-enclosedness and autonomy of the work, we "sacrifice" our own "limited, individual" being to a higher one. The work, then, contains its goal within itself, and this goal is in fact its own completion and perfection (*Volkommenheit*). That the work of art please others, that it receive the public's applause, cannot be the true artist's goal, nor is it the proof of his success; rather, it is only a sign that the artist has achieved his goal within the work. Moreover, the work of art grants us pleasure for its own sake and purely within its own terms; it is not something that can be described or translated into another form or language—unless, perhaps, this description itself be a self-sufficient and beautiful whole issuing from the poetic imagination.[26] And what is the nature of the autonomous work of art's inner completion and perfection? On this issue, Moritz speaks of the necessity of the relations between the parts within the work, an internal necessity that amounts to an external freedom. These internal relations involve the rigorous unity of the whole, a unity achieved, as is the case with all "life, organization, and movement," through the "compulsion of the reluctant parts" to form a whole. For Moritz, this internal coherence is realized through the unification of otherwise conflicting parts within a single tendency and in view of a single goal—this goal and tendency being, in fact, the very state of being together.[27] The key notions here, those requiring an even further and deeper analysis, are those of autonomy, totality, closure, self-perfection, and freedom. One has the impression, in reading these Moritz essays, of the elaborate unraveling of a series of synonyms, for he seems to be saying over and over, in slightly different ways, that the work of art transcends its environment, that it is autonomous in terms of its function, meaning, and mode of organization. This multifaceted emphasis on the work's self-enclosedness and perfection has any number of consequences, not the least of which are the implications for the issue of literary knowledge—be it that of the work or critic. But before such issues are discussed, one point must be made clear: although it would be easy to overlook it, Moritz's theory of the work of art's radical transcendence and autonomy is not a kind of formalism in which the "extrinsic" topics, those pertaining to reception, for example, are simply ignored. On the contrary, his manner of asserting the work's autonomy is based upon

[26]Moritz, *Schriften*, 99.
[27]Moritz, *Schriften*, 47, 49.

a particular conception of the nature of the recipient's aesthetic experience. Again and again Moritz emphasizes the active and constitutive role of the observer, an observer whose "sacrifice" of an instrumental attitude toward the work results in the "recognition" of the work's pure autonomy: "I observe it as . . . ," he notes, going on to say that the observer shifts the goal away from himself and back into the work, where it belongs. Only as a result of an attitude of self-effacement, an attitude of love, can the observer open a space for the work's own transcendence—a transcendence, however, which is nowhere described as the observer's own fiction. Should Moritz say that the work's autonomy depends on its being cognized in this particular manner, that autonomy would vanish in the very same stroke. At the same time, however, he seems to be aware of the fact that he cannot describe an aesthetic experience without discerning at least two poles, those of the work and of the observer—an awareness that brings him to the very brink of stating flatly the constitutive paradox of this kind of aesthetics: the work is transcendent because I sacrifice myself to it; it is autonomous because I make myself dependent.

Here we may begin to perceive some of the mythical elements in Moritz's doctrine, as well as the foibles of the more general tenets with which Todorov associates it. At the heart of both is a single quandary, that of stressing a work's autonomy and independence while also speaking of its relations with what stands outside it. This quandary, which is already quite poignant in the case of the living beings upon which the organicist metaphors were based, is exacerbated terribly in the case of artifacts that clearly only have reality and significance in relation to the human agents who make, perceive, and interpret them. Not to recognize this last point is to embrace some form of idealism (as was the case with Ingarden). Wellek is quite right in hoping to dissociate the aesthetics behind the literary disciplines from what he calls irrelevant "biological parallels," but he does not seem to recognize that to speak of the work as a self-enclosed totality—a totality that we are somehow obliged to respect—does not really amount to a clean break with the idealist background of the organicist tradition. Even less does he perceive the crucial fact that to foreground a certain "multi-layered" and "systematic" mode of "organization" as the crucial features of the literary work of art is to fail to discover anything specific on which to base a separate discipline. This general approach, as well as the specific notions about organization in question here, are simply too similar to those debated within the history of biological thought for us to think of this kind of aesthetics as some autonomous or disciplinary line. (And *that* criticism of the autonomy theory remains wholly within the context of an intellectual and cultural history, not taking the step into a discussion of the institutional dependence of the aesthetic notions.) Should one today wish to take Wellek's suggestion seriously, the most important first step would be to instruct oneself in the literature of those general sciences of "organization," of "multi-leveled hierarchy," and of "systematic totality"—cybernetics, general systems theory, and the theories of self-organization.

Let me return to Todorov's five tenets, with an eye to an even more synthetic statement of the issues. The major quandary, I asserted above, is that of simultaneously asserting the work's autonomy and its dependence. Yet this may not be the most basic pair of terms in which to state the quandary; should we prefer to focus on the hermeneutic or interpretive moment, it would be more crucial to say that the major quandary is that of simultaneously asserting the work's cognitive closure and its openness. In other words, the critic is obliged to commit the heresy of paraphrase, yet this interpretive action remains an act of heresy, doomed to failure; the work's polysemy is inexhaustible, yet the critic is devoted to the hopeless effort of exhausting it. We must experience the work, and we can do so only on our own terms; but strictly speaking, the work is supposed to be closed to us, it is said to be a totality turned inward upon itself, unknowable from the outside. Or perhaps we are more concerned with the issue of finality or function. In that case, the demonstration is to be repeated once more; the major quandary is that of asserting that the work performs no function, that it does not imitate anything outside itself and cannot please or instruct, while, at the same time, we must assert various ideas about what the work does—what it does for the observer, who is elevated by the moment of disinterested contemplation, and what it does in general, as it resolves dualisms and unites dichotomies.

Although Moritz's case may not provide anything approaching an adequate basis for generalizations, the analysis of his aesthetic doctrine does make possible certain hypotheses about the assumptions underwriting prevalent forms of critical interpretation and the "central debate" to which they give rise. The hypothesis can be restated as follows. A critic's emphasis upon the transcendent and self-enclosed status of the work amounts to an orientation that draws the interpreter into an impasse. When the critic implicitly posits the work (or a group of them) as an inexhaustible and transcendent totality of meanings and internal relations, it follows directly that no limited and selective interpretation can ever be valid; and all inquiries are limited and selective. Monroe Beardsley's definition of literature is pertinent here; he speaks of a "discourse in which an important part of the meaning is implicit."[28] What must be added is an injunction to the critic to make explicit this meaning, which is by definition implicit and inexhaustible.

In this light, it is hardly surprising that critics one day discover that this very notion of the work's transcendent totality is also the product of their own interpreting, and that the special status of the work is not the result of its objective properties but of a special attitude having been taken. In this manner we are led to the point where the critic posits the work's absolute closure, its polysemy, and its autonomy as a matter of the purest convention.[29] But this gesture is self-contradictory, because it is the critic who is responsible for looking at the work as a totality. In reality, works of art, texts, or symbolic artifacts are *not* autonomous, perfect, functionless, and meaningful in themselves; they are human artifacts produced under certain circumstances and in specific contexts. To make and display a work of art is never an innocent or pristine act, that is, one lacking a pragmatic and intersubjective dimension, and no work or text can have any meanings whatsoever in the absence of a recipient or observer. The essential truth set forth by the various institutional theories of art is that aesthetic entities are historically contingent and unnatural inventions, and that there probably is no reason to believe in the existence of a transhistorical aesthetic a priori, faculty, need, or "invariant."[30]

My argument against the pure aesthetic attitude under discussion here is *not* simply that it is historically contingent and in a sense fictive, for what follows from this kind of revelation is not evident at all, particularly when one reflects that the same basic truth holds in regard to other conceptions and institutions we would not wish to challenge. Rather, the argument is that this particular aesthetic doctrine asks us to adopt a fundamentally incoherent stance, for we are supposed to respect the transcendence and closure of a work while

[28]Monroe C. Beardsley, *Aesthetics: Problems in the Philosophy of Criticism* (New York: Harcourt, Brace, 1958), 126.

[29]A statement of such conventions is that of Dietrich Meutsch and Siegfried J. Schmidt, "On the Role of Conventions in Understanding Literary Texts," *Poetics*, 14 (1985), 551–74, esp. 556–57. Schmidt and Meutsch think these conventions can actually be followed by competent readers. I criticize this view in my "Théorie littéraire et sciences cognitives," in *Approches de la cognition*, ed. Daniel Andler (Paris: Seuil, forthcoming).

[30]George Dickie, *The Art Circle: A Theory of Art* (New York: Haven, 1984).

any such features can only be the result of our own positing. We are somehow obliged to look at the work as a totality, it being added that the critic should nonetheless try to interpret the work's meanings and study its inner formal relations, both of which are impossible in the absence of any more selective forms of inquiry. Thus the pure aesthetics enjoins us to engage in a process that is by definition hopeless, for the work's meaning was already posited as its transcendence in relation to all contexts, readings, and tasks. Whatever the moral and aesthetic values cultivated by adopting this kind of aesthetic attitude may be, I do not think that we have here a viable recipe for literary critical knowledge.

Patterns of Inquiry

In the previous section I advanced some general ideas concerning certain definitions of the specificity of aesthetic objects and literary works. Although the doctrines in question share the attempt to define the specificity of their objects with the most basic principles of scientific discourse, it appears that there are crippling antinomies inherent in the pure aesthetic concept of the work of art's autonomy. As an application of this concept of autonomy, formalist criticism's apparently scientific insistence on focusing on the specifically literary results in a budget of paradoxes inimical to scientific inquiry as such. Because my discussion of this particular conception is quasi-historical, it is obvious that any strong claims in favor of this hypothesis would require that the theories under discussion be shown to reflect the realities of the emergence and institutionalization of concrete literary research. I am far from considering any such larger, historical hypothesis as being justified or confirmed. Nor is it my intent to argue here for any particular image of the historical emergence of the literary disciplines—such a picture would have to be painted on a much larger canvas and would not be a matter of a purely intellectual perspective on history. My present discussion is in any case animated by other motives, and I make no claim that the extreme versions of pure aesthetics in fact represent the conceptual bases of literary research as such. Another approach needs to be taken in regard to the latter issue. All too often, critical theorists debate over their own fantasies about what critics are and should be doing, and this clash of ideals is not for a second trammeled by the work and the fact of literary research as it manifests itself in countless publications. This weakness of critical theory should be rectified, and it is my goal in the next section to make a small step in that direction. Thus I try to elucidate some assumptions behind recent examples of practical criticism. But what follows is not motivated by any of the more naive empiricist impulses; my sampling of critical work, in addition to being partial (and thus not fully representative), is conducted in function of particular questions. What I propose, then, is an example of the comparative epistemology I have advocated as the type of literary theory needed today. I begin with a discussion of models of explanation in the philosophy of science and then compare some of these very basic explanatory patterns to examples of literary research.

To speak of "scientific explanation" in the singular is to evoke the guiding myth of positivist and logical empiricist philosophies of science, which have invested an enormous amount of effort in the task of arriving at a universally valid model of scientific explanation. For a variety of reasons, post-positivist philosophers of science are often fairly skeptical about the possibility of arriving at such a model—which is not to say, however, that they are therefore committed to relativism, historicism, or sociological reductionist views of science. Rather, the many explanatory successes of the natural sciences are not dependent upon the philosopher's ability to specify the necessary and sufficient conditions of the correctness or adequacy of a scientific explanation. In other words, the rationality of science does not depend upon there being an algorithm for the production or validation of empirical research. Indeed, no rigorous set of such conditions has ever been formulated, and some would contend that there are good reasons to doubt that they ever could be, for it appears that the adequacy of scientific explanations depends upon empirical and pragmatic conditions that cannot be accounted for by any of the universal models of the good scientific explanation.[31] For example, in the classical deductive-nomological account of scientific explanations, an attempt is made to define an explanation's adequacy in terms of conditions that are both logical and empirical. These requirements have been stated as follows:

R1: The explanandum must be entailed by the explanans.
R2: The explanans must contain general laws (or lawlike generalizations).
R3: The explanans must have empirical content (must be testable).
R4: The explanans must be true (if it is to be deemed *the* explanation, and not merely the best one proposed so far).

[31]Peter Achinstein, *The Nature of Explanation* (New York: Oxford University Press, 1983), 137.

This, in a very small nutshell, is a summary of the deductive-nomological model of explanation.[32] It should be noted that statement R2 obscures significant details. Typically, an explanans is described as the conjunction of statements of lawlike generalizations and a description of the antecedent, initial, and boundary conditions of the events to be explained. Roughly put, scientific inquiries typically involve three kinds of terms in this analysis: the particular event or state of affairs to be explained, general patterns (laws), as well as accounts of particular circumstances. To explain, then, is to provide statements of the laws and conditions that can subsume the event. To predict is to foretell the event's occurrence, given the laws and conditions. Induction is the derivation of a law given the conditions and the event.

This type of logical model of explanation, of which I am presenting only a sketch, has been the object of a variety of criticisms.[33] Only a few of them are retained here. A first criticism is that the deductive-nomological model's requirements can be satisfied by extremely trivial examples of "explanations" which, although correct, have more to do with basic reasoning than with science. One must distinguish adequately, then, between scientific and nonscientific explanations. A second criticism is that the model by no means stipulates under which conditions a *specific* instantiation of its pattern is an adequate explanation. First, the fact that the explanans gives *the* operative reason in a given case is not entailed by the fact that the explanans in question satisfies the conditions stated in the model, nor even by the fact that the explanans is true. To make the additional step, other empirical conditions must be satisfied (e.g., we must know that no other causes or reasons explain the explanandum). Even if this problem were to be dealt with and the model could serve as a standard of correctness, the model would still not provide an adequate criterion of the goodness or adequacy of an explanation, which is relative to other conditions (such as the relation between the explanation and the state of scientific knowledge at the time) that require a broader and more pragmatic context of evaluation.

Another major objection comes from the realist camp, which in-

[32]I rely heavily here, and in what follows, on the account given by Willard C. Humphreys, *Anomalies and Scientific Theories* (San Francisco: Freeman, Cooper, 1968), esp. 61–104. This work is rare in its able combining of formal and historical perspectives. For background to the anomaly conception, see Carl G. Hempel, *Philosophy of Natural Science* (Englewood Cliffs, N.J.: Prentice-Hall, 1966).

[33]For a clear and detailed presentation, see Achinstein, *The Nature of Explanation*, as well as Craig Dilworth, *Scientific Progress* (Dordrecht: Reidel, 1981).

sists that science cannot explain why events happen if the so-called explanation is all purely a matter of the logical relations between its own statements. Missing is the constitutive or etiological basis of a real explanation. Many of the important scientific explanations, then, are not sufficiently described in terms of a logical subsumption of an event beneath a more general statement; the explanation also involves a reference to some mechanism or process that is responsible for the regularity plotted in the scientific statements. One example used to illuminate the weakness of the purely logical model of explanation is that of the flagpole and its shadow. Given the possibility of providing lawlike statements that describe the regular relations between a flagpole and its shadow, what, in the deductive-nomological model, prevents us from saying that the height of the flagpole can be explained in function of the length of its shadow?[34] It may be possible to deduce the length of the pole from that of the shadow, but it is not the shadow that is responsible for the length. A scientific explanation must say something about *why* the length is what it is. According to some philosophers, what is missing in the formal account, then, is an idea of causality, or of a natural or material necessitation, which is not reducible to a mere regularity of relations.[35] Thus the realist objection requires the addition of a fifth requirement, which can be roughly stated in the following way in our context:

R5: The explanans must be interpreted as referring to a productive or causal process.[36]

[34]This argument was first articulated in these terms by Sylvain Bromberger, "Why-Questions," in *Mind and Cosmos*, ed. Robert G. Colodny (Pittsburgh: University of Pittsburgh Press, 1966), 86–111; for a more recent discussion, see Clark Glymour, "Two Flagpoles Are More Paradoxical than One," *Philosophy of Science*, 45 (1978), 118–19.

[35]A particularly clear statement of this argument is that of Wesley C. Salmon in "Comets, Pollen and Dreams: Some Reflections on Scientific Explanation," in *What? Where? When? Why?: Essays on Induction, Space and Time, Explanation*, ed. Robert McLaughlin (Dordrecht: Reidel, 1982), 155–78. Salmon gives examples of scientific explanations that relate noncausal regularities to underlying causal processes. Whether this fifth stipulation is too stringent is an issue that cannot be taken on in detail here; the answer has to do with the status one wishes to grant to probabilistic laws within specific domains of scientific knowledge, their relation to more causal nomic generalizations, and, of course, the definition of causality in question. On causality, one may usefully refer to Michael Scriven, "The Logic of Cause," *Theory and Decision*, 2 (1972), 49–66; and Raymond Martin, "Singular Causal Explanations," *Theory and Decision*, 2 (1972), 221–37.

[36]Here I insert a stipulation that does not figure in Humphreys's account. He seeks to deal with the flagpole example with a contextual limitation on the anomalies entertained—an important point that I have introduced below in R8; see Humphreys, *Anomalies*, 80.

The addition of this kind of requirement has itself drawn criticism from various sides, for it is pointed out that not all of the explanations that are presently considered examples of good science involve reference to causality, at least under any of the major competing definitions of the latter. If to explain is a matter of giving reasons why some state of affairs holds, causal reasons are only one kind of reason. Moreover, it is notoriously difficult to give a rigorous account of causal relations. Perhaps the important realist intuition behind R5 will eventually be salvaged by means of a formulation capable of embracing this diversity of explanatory modes while isolating the specificity of those explanations that refer to a constitutive or productive natural (or social) mechanism.

Another major objection to the deductive-nomological account arises from a basic insight into the selective and oriented nature of scientific inquiry; science does not seek to explain *everything*. It is, of course, the Laplacean dream to be able to subsume every event or state of affairs beneath a covering law and set of initial conditions, to be able, as a result, to predict and retrodict everything. Actual scientific inquiry, however, is a social and human activity, not a state of divine omniscience, and thus, even if Laplace's model were taken to be its vanishing point or regulative Idea, its manner of advancing toward that infinitely remote goal could not be as simple as the deductive model suggests. The question that arises is this: given the finitude of scientific knowledge, *which* events and states of affairs should it seek to explain? As the most basic features of the deductive model suggest, a first and broad answer to this question is already available: insofar as science seeks constantly to relate the individual states of affairs and events to general patterns, types, laws, and necessarily recurrent conditions, it seeks to explain only those aspects of reality that are repeated. But this is only a first and wholly general principle of selectivity behind scientific inquiry, for what is more crucial is to see how the selective nature of explanations is manifested in a much more specific way. This is the thrust of the "anomaly" theory, which has been set forth by Willard Humphreys. In this account, a scientific explanation is not simply any chain of reasoning having the deductive-nomological form; rather, it is one whereby an anomalous state of affairs is at stake, one in which the explanandum is this anomaly. Thus one may add yet another requirement to the account of explanation being drawn:

R6: The explanandum statement must be the description of an anomalous state of affairs or events.

But this requirement hardly settles the matter, for it remains to be seen how one may decide which occurrences or states of affairs stand forth as real anomalies. After all, I can look at anything at all and consider it to be strange and in need of explaining. The recognition of anomalies is relative to a context of belief. It is only in relation to a body of beliefs, a theory, for example, that something stands in need of explanation. Thus a truly anomalous situation is one in which certain expectations, following from a given set of beliefs, are overturned by what appears to happen, so that the beliefs and events need to be reconciled within a more coherent scheme. It may be necessary to change the conception of the laws or patterns governing such situations; it may be necessary to alter the boundary conditions and statements of antecedent conditions; or perhaps the event can be interpreted in such a way that it no longer stands as an exception to the rule. In any case, one recipe for an explanation here adds another general requirement:

R7: The explanans must solve a conceptual problem posed by the context of beliefs.

This requirement, however, is not good enough, for the number of anomalies that can be generated in this manner is still far too large to account for the selectivity of science. It would be possible still to conceive of a host of anomalies relative to any number of beliefs, those involved in wild counterfactuals, or cases of extreme ignorance, for example. Humphreys, who is primarily concerned with the natural sciences, accounts for this problem by stipulating that the anomaly must be a natural one. This seems overly restricted insofar as science need not only be thought to explain natural realities. Moreover, the question remains concerning *which* natural realities should be deemed important anomalies. Another of Humphreys's formulations suggests the way to improve the stipulation; he suggests that the statement of affairs described as a "natural" anomaly is one that contradicts an ensemble of statements presently held to be true.[37] Yet this is still too loose. Held to be true for what reasons? Surely it is the current state of scientific knowledge that must be left holding the reins here, so a rough approximation of the requisite stipulation is this:

R8: The explanandum statement must be the description of an anomaly relative to science's best available theory, for example, those

[37]Humphreys, *Anomalies*, 7–8.

> beliefs presently held to be true in that theory. The explanans must include as many of the assumptions and beliefs of the theory as is consistent with the other requirements; there must be a minimal deviation from the context of beliefs in which the original request for an explanation is cast.

Even this last stipulation probably leaves us with a highly idealized view of scientific inquiry. First, it evokes the image of a monolithic Science instead of a multiplicity of research programs. Second, it leaves out of the picture the ways in which science's general selectivity is further narrowed by exogenous factors. The diverse social institutions within which research is conducted cannot possibly be invested in explaining all of the anomalies that could appear relative to the best theories in its different research programs; clearly there are other, more specific research priorities involved in singling out anomalies and determining which will be worked on. Many of these priorities are motivated by commercial, military, and political interests, and such matters must be taken into account in an adequate conception of science.

It should then be clear that we have not yet arrived at anything remotely resembling a set of necessary and sufficient truth conditions of reason-giving propositions, nor a binding model of the criteria for a good explanation. Nor, however, do we need such a perfect model if our goal is an epistemological critique of literary research. In spite of their limitations and points of contention, the discussions of explanation I have touched upon point the way toward a conception that may usefully be brought to bear on the case of literary critical inquiry. Although the philosophy of science cannot offer a list of universally valid instructions for the production of good scientific explanations, its current findings can, in a broad way, be used to shed light on the characteristics of explanations in other fields. For this purpose, an open-ended and incomplete model is wholly adequate. Thus, in what follows I adopt the following, broad conception of explanation: to explain is to give reasons why a given state of affairs obtains; this providing of reasons is achieved by relating the particular case under explanation to a more general pattern. It is the general pattern that is held to explain why this state of affairs is the case. An explanans, then, is an account (or even a sketch) of a pattern, on the one hand, and the particular circumstances in which it is manifested, on the other. Sometimes an explanation is a matter of subsuming a particular case beneath a previously established pattern or law, but scientific progress more frequently involves the discovery of laws or patterns capable of plotting significant variables isolated within complex systems. In this broad conception of explanations, no restrictive definition is provided of the explanatory relation that links a particular case to a general pattern or law (e.g., I am not assuming that all explanations must be causal). It is assumed that the explanandum or state of affairs to be explained is defined in the terms of a given body of beliefs about the domain under consideration. In the case of scientific explanations, those beliefs are ideally representative of the best available knowledge—knowledge amply supported by reasons and evidence. Failing the satisfaction of this "best available knowledge" clause, an anomaly can still be singled out in relation to one of the major research programs competing within a field. In either case, it is the available knowledge that initially specifies the particular type of explanatory relation that is posited between the pattern (law) and particular explanandum. In this simplified and broad notion of explanations, the "nomic" requirement is being considerably loosened in relation to the deductive-nomological model. This step is warranted when historical and cultural domains are under examination, for here, the idea of rigorous general laws (involving strict universality and necessity) is not readily brought to bear, and one is more likely to speak of lawlike generalities, regularities, or patterns.[38] It may be noted that in the broad notion of explanation just outlined, the distinction between scientific and nonscientific explanations (both of which count as knowledge) is not sharp and involves varying degrees to which certain methodological norms are satisfied—these norms involving such notions as precision, simplicity, and scope of application. This absence of any sharp distinction between scientific and other forms of explanation is not a problem in the context of a discussion of explanation in areas where there is a variety of underdeveloped research programs, and where the important issue is the difference between explanations *tout court* and nonexplanations. The basic features of explanation outlined

[38]This sudden loosening of the "law" requirement may raise some eyebrows, but I do not think that such a move in fact warrants the typical skepticisms about explanation in the historical and cultural domains, nor the old conclusions about idiography. What is ignored is that the model of lawlike explanation is already an idealization in relation to the very best explanations provided within the natural sciences; one should read the literature on lawlike explanations in history while keeping in mind David Bohm's comment that "there is no known case of a causal law that is completely free from dependence on contingencies that are introduced from outside the context treated by the law in question"; in his *Causality and Chance in Modern Physics* (London: Routledge & Kegan Paul, 1957), 61. See R. F. Atkinson, *Knowledge and Explanation in History* (Ithaca: Cornell University Press, 1978), esp. 102–15.

above are pertinent to this latter difference, as are the most basic principles concerning the difference between knowledge and mythology, inquiry and dogmatism.

Literary Explanations

I shall now examine some aspects of literary-critical research with the previous discussion in view. First a liminal observation, which may be in no way essential. The conceptual problematics of many literary-critical publications are highly implicit; nowhere is there provided, in many such publications, a specific statement of the questions guiding the research, and even less frequently are any of the specific background assumptions or beliefs made clear. The discourse begins and ends with a flow of assertions, uninterrupted by any explicit metalevel statements about the ambitions and limits of the effort under way. One may ask in this regard whether literary publications differ essentially from those in other disciplines; perhaps the theoretical frameworks and issues that guide particular contributions are on the whole taken to be a knowledge common to the insiders, a knowledge that may therefore conveniently be left in the background. Or perhaps the difference is largely rhetorical: on the one hand, a penchant for the explicit, on the other, a preference for the more suggestive and nuanced style of the literary essay. But in any case, the question whether an understanding of the nature of the problematic at hand is really shared by the readers of literary essays, even the minority of expert readers, should not simply be avoided. In my own case, I must confess that to come to some understanding of the purported assumptions and aims of certain literary essays requires a certain amount of real guesswork, not because the question of author's intent is always a thorny one, but because it is simply not made clear what questions the particular piece of work is seeking to address, what background assumptions are at work in the formulation of the issues, and what bodies of knowledge and priorities are relied upon in the process of proposing an answer. Moreover, there is frequently a real dearth of explicit *methodological* thought; when questions are posed instead of assertions simply made, there is often a lack of any explicit metalevel discussion of how such questions may or may not be addressed, what would and would not count as telling answers, and what evidentiary norms are held to be applicable. In short, at first glance it seems that many of the published findings of literary research do not appear to satisfy many of the most basic norms and principles of research. In the place of clear and well-delimited hypotheses, posed as such, is a stream of assertions frequently couched in a language in which no effort is made to define central terms, and in which questions of testing, confirmation, and justification are left entirely up in the air.[39]

My next observation concerns what surely must be called an anomaly within my own present conception; whereas, on the one hand, I lend credence to a view in which research is understood as a problem-solving, oriented, interrogative activity, an activity that seeks to propose answers to questions, on the other hand, in many literary publications there is not a trace, explicit *or* implicit, of any such effort to resolve a "cognitive tension." Let me call this class of critical publications, a class that is, I think, not small, "megaphone criticism." By this I mean to say that the critic appears to understand his or her task as a kind of amplificatory restatement of the message or vision of the works, authors, or genres in question. Let me give a slightly more precise description of "megaphone criticism." In a typical example, a critic singles out a novel by a particular author, usually a figure falling squarely within the canon as it is circumscribed by a general consensus. Furthermore, the critic singles out a theme, topic, or issue that is dealt with, explicitly or implicitly, within the novel, the author's other texts, or both. The critical essay then sets forth as coherent and detailed a picture as possible of what these utterances have to say about the topic in question, citing evidence and so on. Thus we have a basic pattern of explanation in which the central task of the critic is to make explicit a literary message by answering the following implicit question:

Q: What did literary text *x* say about topic *y*?

Frequently the reader is nowhere told such things as (1) why this procedure of elucidation is necessary, (2) why it is important or valuable, or (3) what the critic thinks about the relation between the amplified message's version of the theme and the truth of the matter. The reader is not told why the message is valuable as a means of correcting some other, inadequate doctrine, or why it would be wrong to be led astray by its apparent charms, for example. On the contrary, it seems to be taken for granted that a detailed and coherent

[39]For clear statements of many useful and basic methodological norms that need to be followed more carefully in literary research, see the works of Siegfried J. Schmidt; for example, "Literary Science as a Science of Argument," *New Literary History*, 7 (1975), 467–81.

restatement of the messages that can be derived from the symbolic artifacts in the literary canon (construed very broadly) is an end in itself. The only truth claims relevant to this kind of criticism seem to concern what the message is, not the validity of the message itself. Nor does the critic ask why this particular literary text was generated. Getting the messages right therefore constitutes a research contribution.

Clearly what we have in this type of example is a way of construing research that is far looser than the previously outlined model of explanatory inquiry, so very loose that I am tempted to resolve my own "anomaly" by concluding that the class of examples under consideration is not a matter of explanatory inquiry at all. Only were we to grant the literary discipline a perfect autonomy within the institutions of knowledge could we resist this conclusion, for only then could we allow that discipline to establish its own norms of what constitutes successful research—norms that seem to fly in the face of the standards entertained elsewhere. But this conclusion may be overly hasty, for we have failed to get at the implicit assumptions that make the megaphone approach to research more plausible. Why does the critic not deem it necessary or appropriate to contrast the elucidated message's implicit validity claims with some other body of knowledge about what is really the case? A highly plausible response is the following: the critic's most basic assumption is that what he or she is interpreting are not, first and foremost, earnest attempts to state the truth of this or that state of affairs but literary artifacts that, although they are held to be highly meaningful, are not to be approached in search of that sort of validity. The work is an oracle to be interpreted, but the oracle finds its place within the brackets of fiction. To make the accuracy or inaccuracy of the message the explicit and guiding goal of criticism would be to abandon this aesthetic premise, which seems to dictate that the only validity in question here is that of the critic's own account of the message. The megaphone may be faithful or unfaithful, but it is wrong to apply such terms to the message it seeks to amplify.

Before any of the implications of these latter points are explored, it must be noted that megaphone criticism is not always so straightforward. Frequently an element of cognitive tension, a first type of explanatory problematic, is provided by the existence of rival accounts of the message. In this manner the research begins to resemble an explanatory activity; the critic establishes a background of critical opinion, contrasts this opinion to evidence, and resolves this tension in favor of his or her own opinion as to the nature of the message. Thus literary explanations take the following form: critical opinion says *p*, additional evidence suggests not-*p*, but *q*. This argument structure is not just a rhetorical veil, for it partakes directly of the most rudimentary pattern of inquiry.[40] This latter point becomes all the more telling when one looks at the different types of arguments that are elaborated within that general structure, for they too obey a basic explanatory model insofar as the general principles are concerned. In a moment I point to some types of arguments as illustrations and examples. First, the most basic structure, which is the literary explanatory pattern produced in response to inquiries of type Q:

> LE: Critical opinion on the meaning of a given literary phenomenon presents one or more coherent conceptions, but none of these has accounted for certain additional evidence. A more comprehensive and coherent account, fitting this additional evidence into a pattern, is produced.

This is, of course, only the baldest transcription of the anomaly view of explanation. In the following I present more details based on an examination of actual pieces of literary research. What I am out to provide is a suggestive list of different kinds of interpretive arguments, not an exhaustive tableau of the critical *topoi* or any empirically correct statements about relative frequencies. Because my goal here is to suggest a fruitful path of analysis and not to present definitive conclusions, I think it unnecessary to identify the samples upon which my remarks are based. Here then are some argument sketches based on pieces of literary research recently published in a variety of major journals in the field:

1. The meaning of a passage in a work is particularly unclear, and thus anomalous in relation to the message that is clearly evidenced in the rest of the work. By going outside the work to other evidence relative to the life, corpus, sources, genre, or tradition, the sense of the passage can be amplified and seen to fit within the overall message.
2. The meaning of a work is unclear, and critics disagree. By understanding it as an example of a given genre, some major features of its message are made clear, and the disagreement is settled.

[40]My point here is hardly original, and many critical theorists have perceived a *basic* similarity between scientific and critical reasoning. For example, the idea is forwarded by John Ellis, *The Theory of Literary Criticism* (Berkeley: University of California Press, 1973), 190. Ellis's reliance upon Kuhn in the same context, however, raises many questions.

3. A work clearly belongs to genre *x*, and works in this genre do not typically convey messages of a certain type. Yet an established account of author *y* of work *x* suggests that this author would want to convey a message of that type. A closer look at the evidence of the work and of the author's attitudes reveals that certain of the genre conventions are in fact violated so that a message uncharacteristic of the genre is subtley conveyed.
4. The meaning of work *x* is unclear because of the difficulty of understanding what appears to be a particular type of ironic voice. Given a broad taxonomy of ironic discourses, the variety of *x*'s irony can be situated and understood and the meaning of the work clarified.
5. The message of author *x* on topic *y* is *p*; the message of author *q* on topic *y* is not-*p*. Relative to a coherent account of both authors' general aesthetic (or other) attitudes, their views are usually congruent. To explain, then, the existence of both *p* and not-*p*, one must either reject the view of congruence or its bases, or identify the circumstances behind the disagreement.
6. An author's message on topic *x* in work *y* is understood one way when viewed in the context of trends characteristic of one national literature's history. That message does not, however, match a fuller account of the author's perspective. The message of work *y* can be understood quite differently when seen within the pattern of that author's own corpus, and thus the meaning of *y* is settled by deciding in favor of the latter context.
7. An author's corpus has a coherent series of meanings, except for work *x*. Work *x* can be more fully integrated into the corpus by altering the account of the overall pattern of the message's meanings, or by adducing circumstances under which the pattern is not broken by the existence of *x*.
8. Critics contend that there is a major discontinuity in author *x*'s corpus and worldview, a change that occurred in midcareer. But this is wrong, for additional evidence—particularly proof relevant to the date of a crucial influence—shows that the change occurred much earlier.
9. An author's message is ambiguous and could even be contradictory and incoherent. As the author is recognized as great, critics are hesitant to settle upon such a judgment. But they should do so, because an awareness of the disparate materials and traditions that the author had to bring together explains why even this author could not make them fit together coherently.
10. The message an author conveyed in one group of works is less forcefully conveyed in other works by the same author. The discrepancy is explained by the difference in genre.
11. Great author *x* composes work *y*, which lacks originality and appears to present only the banal message typical of the genre to

which it belongs. But a closer look at *y* shows that this view of *y* is wrong.

12. Author *x* appears to be saying *y* in the corpus of *x*'s works. However, *y* is relative to a superficial theory of meaning; once inserted into the context of a deeper and more comprehensive theory of meaning, the message is correctly understood.
13. Critics have assumed that work *x* has a coherent message. Evidence is used to show to that it does not. This example of an incoherent masterpiece is further evidence of the general pattern, which is that literary messages are not really coherent.
14. It is held that all important literary works deal with a given theme (writing), and it appears that work *x*, which is usually thought important, does not. Yet a closer look at the evidence reveals subtle metaphors associated with writing in *x*.
15. All works of art convey messages relevant to unconscious processes as identified in theory *x*. Work *y* conveys such messages and hence confirms the pattern.
16. All works of art convey messages relevant to unconscious processes. Theory *x* identifies those processes in one way, theory *y* in another. Work *q* conveys messages that are not those specified in theory *x*, but which do fit those of theory *y*. Theory *y* thus gains additional support.

Although these blank descriptions have the shortcoming of stripping the actual examples of the wealth of learning and subtlety which they almost always manifest when seen "on the hoof," the analytical skeleton does reveal the presence of a certain invariant pattern of inquiry. Almost always, we find that the critic has in mind some account of the circumstances and a pattern but moves on to point to the existence of overlooked evidence—thereby identifying an anomaly. The claim is that the evidence does not fit within this pattern in any obvious way, and thus a new account should be sought. And this is what any number of critics in fact do, often with admirable diligence, skill, and creativity.

But if many critics are at least implicitly engaged in work that approximates a very basic model of inquiry, how is it that they produce such divergent and apparently noncumulative results? The patterns of literary explanation seem to be made and broken without there being anything remotely resembling overall progress. Why is this so? An obvious answer points to the notions of what amounts to a good account of the circumstances and patterns to be correlated with a symbolic artifact's meaning. Insofar as these assumptions differ significantly, the ensuant explanations also vary wildly. It is also not the least bit obvious what kinds of evidence are and are not to be included as determinations of the artifact in question. The choice of the units of analysis, and of the variables and properties to be isolated in them, is not clear. We are back to the indefinite nature of the question being posed: What does the pendulum mean? In other words, the admirable explanatory efforts of a large part of interpretive criticism fall within the scope defined by Q, and a first and large source of ambiguity resides in the indeterminacy of the notion of a literary text that figures there.

What is obvious, then, is that the debate has to move to another level if it is to be productive at all, that is, if there is going to be anything like an understanding of the possibility or impossibility of validity in interpretation. Thus the problem of the coexistence of different megaphones that always end up broadcasting different messages leads to the search for some kind of hierarchy of amplificatory techniques. We are led back to what I have called the central debate. What is wrong with this debate, I want to argue, is that it remains fully within the bounds established by the assumptions that orient megaphone criticism as a type of inquiry in the first place. For even when critical inquiry and debate strongly resemble a basic logic of explanation, with its processes of cognitive tension and resolution—as in the types of cases outlined above—the essential first step is still one that assigns to critical inquiry a fundamentally amplificatory role. The critical debate is ultimately limited and sterile because the kind of veracity that could ever be sought within its terms is profoundly limited from the start: critics can argue as earnestly as they like over the respective validities of their statements about the messages; as the central debate becomes more and more central, they can argue over the validity of their approaches and methods in identifying messages; but they cannot assert or deny the validity of the messages themselves without stepping outside the bounds of the discipline, at least as this discipline's specificities have generally been defined. Moreover, it can be argued that this same first step, which places a ban on claims concerning the truth and falsehood of utterances that are at once meaningful and artistic, is at least one of the major determinants of the chaotic nature of the results obtained within the field of inquiry it defines. The critics are required, at least following this particular orientation of the discipline, to ask and answer an indefinite question. This does not mean that they can ask and answer any sort of question at all; the variety of their inquiries and approaches is overshadowed in advance by the specific nature of *the* indefinite question that guides and misguides, orients and disorients, the inquiry. In other words, and more bluntly, "Let us find out what the message of the artist or artistic creation is" is not a neutral orientation of research. Nor is it precise. Moreover, in certain of its applications, it is flatly self-contradictory. This is the case whenever it is assumed that the meaning to be excavated in critical research is an inexhaustible *je ne sais quoi*, which could never be fully stated.

To recapitulate, I have argued that critical research sometimes manifests a rudimentary explanatory pattern that is similar, in its most basic form, to scientific explanations. Yet I have also argued that this research, and the various local explanatory efforts within it, are often based upon a first assumption or orientation that severely undercuts and restricts the explanatory nature of the results—so much so that to speak of "explanations" would in many instances be erroneous. Locally, these critics seek to provide explanations, and they do so with admirable rigor, industry, and invention. Yet globally, these research efforts are not part of an explanatory structure at all. Locally, the literary explanations are a matter of a long and difficult process whereby patterns and circumstances are hypothesized, investigated, and described and evidence is brought into confrontation with these patterns. But globally, the question being posed, in myriad different forms, does not arise as the result of a precise cognitive tension between a well-founded theory and the states of affairs that it has yet to account for. What theory or context of belief is in question if all that is asked for is a series of coherent accounts of the literary texts of the past?

Here, of course, I may anticipate an obvious objection, the frailties of which are easy to reveal. How was it determined, one may protest, that literary critics should be trying to *explain* anything in the first place? Is not their properly humanistic task that of *understanding*, not explanation? What is wrong with this objection, and with the general approach it defends, is the fuzziness and incoherence of the notion of understanding it advocates, for insofar as such understanding is truly different from explanation it does not make a good candidate for knowledge in any strong, systematic sense of the word. It is granted, of course, that readers can and should "make meaning" out of literary texts; it would be a flagrant misreading of my perspective to think that I do not consider the interpretation and appreciation of literary works to be a valuable activity. But it is another question whether such activities should be considered genuine research. Surely the fact that critics can "make meanings" and engage in conversations is not a sufficient answer to the question of the status of literary knowledge. Can we really speak of an alternative form of knowledge, one that can be set in opposition to explanation, when the results of these understandings are a discontinuous and noncumulative jumble—the very jumble that has led to the emergence of the central debate and to incessant talk of the crisis of the discipline?

Not only is it wrong to pretend that everything is in order within the unbridled interpretative practices, but it is also important to reflect upon the other types of inquiries that are precluded as long as these practices continue to preoccupy the majority of literary scholars. This issue provides the focus of the next chapter, but here we may broach a single question along these lines: at what point does the goal of producing these various understandings of texts, viewed as works of art, preclude the explanation of the historical emergence and consequences of literary utterances? The distinction between texts and utterances is crucial in this regard, for while the former refers to the artifact of a particular type of semantic representation, the latter term embraces the pragmatic dimension, or, in other words, the historical contexts within which literary messages are created and received. Could it not be the case that the massive indeterminacy of the literary messages, which has proved to be the stumbling block of literary research and its central debate, is ultimately the artifact of a first step whereby the utterance is viewed as a kind of sacred source of meaning?

Surely if this particular type of literary knowledge is not going to be given as an end in itself, those who wish to define and defend it must begin by proposing strong answers to the kinds of questions that have just been raised. All too frequently, megaphone criticism, as well as its theoretical overseers, simply beg these issues; they are too concerned with asking whether and how it is possible to get the message right, and consequently they fail to ask why this approach to messages should be adopted in the first place.

It should now be possible to trace the lines of inquiry that proceed outward from this point in my argument. I have argued that, in a large portion of critical practice, the critic's sense of the discipline's specificity is implicitly defined in terms of a particular approach to the interpretation and analysis of literary texts, this approach being, at a very basic and general level, that of Q. This general question of the message in the infinite bottle of literary texts surely admits of any number of answers, yet to speak of answers here is rather dubious insofar as the issue of their global status is concerned. For this reason, it is crucial to examine at much greater length the limitations of Q as a basic first step. This examination should take up, in my view, several related issues. It must seek, most of all, to understand Q in relation to the other possible questions that may guide research within the historical, social, and cultural domains. In this manner, the alternatives to the literary focus can be discerned, and it should be possible to conceive of a reintegration of literary investigations within a broader, interdisciplinary framework where more precise and realistic explanatory questions may be approached. At the same time, what is needed is a probing analysis of the assumptions that have implicitly guided the literary discipline's fixation upon specific forms of Q. I stated in passing that these assumptions were at once of an aesthetic and a historical nature, and some excellent work has already been done on this score, work that indeed goes far toward revealing the contingent historical processes whereby the fiction of a certain form of literariness was established as a dominant institutional fact or social-imaginary signification. The inherent, and in my view crippling, limitations of Q are best revealed, however, by comparing this type of query to other possble questions. To show that a certain set of aesthetic beliefs is historically contingent does not, in any case, amount to establishing that it is wrong; just because a purely aesthetic definition of "the literary" did not emerge until the end of the eighteenth century is not in itself reason for us to reject this definition, even if such an historical perspective has the real merit of challenging any view whereby this definition is held to be simply self-evident or obviously universal. Thus in the next chapter I attempt an epistemological analysis that draws distinctions between some of the various types of questions that can guide the interpretive endeavors of criticism, and I propose two alternatives to the sterile game of the message in the bottle.

Thus I am exhorting literary critics and their students to devote less time and energy to the kind of interpretive practice I have been discussing. Moreover, I believe that the theories that debate the possibility of finding real knowledge *within* that type of practice are futile. In this sense, I partly agree with the opinion of such theorists as K. M. Newton who contend that critical interpretations can only have the validity granted them by the institution, which embraces a range of different assumptions and perspectives in an overall balance.[41] But I do not think that such judgments are warranted in

[41]K. M. Newton, "Validity in Interpretation and the Literary Institution," *British Journal of Aesthetics*, 25 (1985), 207–19.

relation to the full range of possible interpretive strategies, and I would like to see an upsetting of the present institutional balance in favor of interpretive research aimed at a different kind of objective. It seems curious to me that at the moment when critics and theorists are becoming increasingly aware of the contingent nature of the critical institution and its norms, they should elevate the present framework as a final authority.

CHAPTER 7

Hypotheses Fingo

I argued in the last chapter that the assumptions underlying some predominant types of literary research require a thorough reexamination. Whereas various distinct avenues of inquiry are open to literary critics, the differences between the goals and assumptions of different kinds of work are frequently obscured, the result sometimes being a failure to pursue any of these aims well. In this chapter I discuss the relations between some of criticism's possible orientations, with particular reference to the place of literary knowledge within the anthropological disciplines.

One of my objections to a certain vein of literary research is that its explanatory efforts are crippled by the initial orientation, an orientation that sets for the critic what is by definition a highly indeterminate task; the critic is led to the seashore and is asked to find *the* grain of sand. What, more precisely, is the nature of this orientation, which places the critic in an essentially aporetic situation and leads the theorist into the impasse of the central debate over validity in interpretation? In the last chapter I contended that a first and very fundamental feature of this orientation amounts to criticism's tendency to pose a wholly global question concerning the meaning of a literary work. I have argued that this is a terribly indefinite question, and in fact no one can tackle it as it is. Instead, the critical observer or reader of a literary text implicitly or explicitly takes on a different

For background to Newton's famous slogan *hypotheses non fingo* ("I make no hypotheses"), see Willard C. Humphreys, *Anomalies and Scientific Theories* (San Francisco: Freeman, Cooper, 1968), chap. 1.

question, asking what the text means in regard to some more or less well-defined context or background. Thus the vague question that typically orients and disorients literary research amounts to a vague and indefinite avenue of inquiry, one the critic must struggle to disambiguate. Yet for all its indeterminacy, the literary approach *is* characterized by certain very basic assumptions. Although I have said nothing rigorous or detailed about these assumptions, it is possible at this point to note that the literary approach is a matter of isolating, from the ocean of historical realities, a series of symbolic artifacts and agents that are identified as special and particularly valuable sources of meaning. A key feature of the critical attitude is the desire to interpret these meanings, to paraphrase, amplify, and broadcast them anew. Much more can be said about the different ways this broadcasting is done, but at this point it is equally if not more important to say what is *not* done in this style of research. What is not done with such messages has to do with an essential premise about aesthetic communication, one following which the artist or work of art speak "within brackets"—and hence produce utterances that should not be assumed to be true or false. Nor are these utterances, according to another deep-seated aesthetic premise, to be evaluated in terms of their immediate utility. To view literary utterances from the optic of a theoretical–practical framework is the very essence of the Philistine response. In matters artistic, we would do well, it is held, to adopt the kind of ceremonial attitude recommended by Wittgenstein in regard to ritual.

But what am I enjoining critics to do instead? The only point upon which I insist in this regard is that there is a wide range of alternatives. I make no effort to provide any exhaustive list of these other options. What I undertake is to isolate and promote two distinct avenues of inquiry which in my view deserve more attention. The first to be discussed concerns the possibility of a reintegration of the study of literary works within a realist sociohistorical framework of analysis. Although it is hardly new to condemn the centrality of aesthetics in the literary fields and to call for historical or even materialist work, my proposal differs from some manifestations of these tendencies. I seek some guidelines for telling whether the critic has really got beyond the aesthetic a priori, for I think that to do so involves a much more basic shift than many critics realize, the result being that many putatively sociological approaches to literature are still vitiated by aspects of a rather different orientation. My second alternative concerns a more properly epistemological and cognitive role for literature and its analysis and calls for a reintegration of literary work within the ongoing project of hypothesis formation within the human and social sciences. I take up these two alternatives in turn.

From Messages to Consequences

In this section I contrast some of the typical aesthetic assumptions behind literary readings to a simple model of communicative interaction, my goal being to bring out some of the most striking differences. I want to insist that my aim here is not to *reduce* the broad range of literary practices to this model—far from it. All that I am proposing is a kind of thought experiment that serves to highlight the differences between *one kind* of historical research and the selective orientation that frequently characterizes sociological and historical orientations to literature. Readers who think that a sociohistorical analysis of literary texts is not pertinent to their field are invited to skip to the next section.

Let us imagine, then, a correspondence between two parties. The domain to be modeled here is an actual exchange of letters, that is, one occurring between specific human agents in some specific social and historical context, an exchange in which both participants share a specific form of language with its particular norms and practices, rules and conventions, tacit and explicit knowledge. This domain is enormously complex, and no one could ever hope to describe or explain it exhaustively. But we want to make some maps of it that single out some crucial and determinant features. This exchange of letters is not to be thought of as some sort of abstract sending, receiving, encoding, and decoding of messages—as in the usual bloodless and abstract models of communication. The situation is not that of two simple machines coupled together in a rudimentary feedback system. Rather, it is important to stress the historical and social nature of the context in which the exchange of information takes place, and it is also crucial to see that the idea of sending and receiving information is far too abstract to give an accurate sense of what is going on. Although it seems a good idea to agree that the two correspondents are sending pieces of paper back and forth to each other, let us not leap to the conclusion that what they are engaged in can necessarily be described in similar terms, as if they were sending parcels of meaning back and forth as well. Instead, let us try to imagine that some kind of transaction is taking place between our two parties; let us suppose, for example, that in these letters they engage in a discussion over various issues and states of affairs, and that this discussion is not just a series of referrings but an interactive process in which their relationship to each other is being reproduced and redefined. Here I am only adopting the most basic pragmatic intuition, which notes that the function of language is not purely referential, and that the other, varied functions of symbolic action cannot be simply derived from the signs' reference. But that does not mean that the letters do not refer to anything. The issues taken up may concern the writers' relation to each other; they may concern other persons known to them both, or more general states of affairs or notions. And let us not forget that their correspondence may include various forms of interaction, not only attempts to write what they think in a full, honest, true, and unambiguous way; the transaction may also include jokes, ironies, ambiguous and veiled statements, a concealment of intent and belief, a repetition of highly ritualized or habitual formulas, and so on. No doubt this example is still bloodless and abstract in relation to the complexity of any actual interaction between two people. But to simplify, let us call one of the agents A, the other B, and let us extract from the flow of their exchange one of the topics of their correspondence, which we call *x*. Abstractly, *x* could be any state of affairs. To give an example, we might suppose that *x* is a marker for the fact that a friend of theirs is unemployed and that this unemployment has become a dreadful problem for this friend, a problem about which they both may be concerned, and over which they exchange observations. This particular topic may have emerged in a relatively spontaneous manner; it need not be that one of them wrote specifically in order to deal with this issue, and they may not be exchanging views with the goal of making any resolutions or of settling on any course of action. They may or may not, for example, have the intention of revealing to the friend their thoughts on the matter. (But this particular version of *x* is only an example, and very different ones could be given instead.)

Let us now engage in one more abstraction. Let us imagine that we can extract from this exchange of letters a particular missive written by A. We have selected this missive because it is the one in which A broaches and discusses the topic of *x*. Here add the convenient assumption that A does not discuss *x* in any of the other letters. In B's next letter, B responds with some comments of her own on the topic of *x*, but A does not return to this issue in the following letter. In other words, the "internal evidence" (in the narrow sense) pertaining to what A writes about *x* is entirely exhausted by one letter, which we call L. To simplify the model, assume that none of the difficult empirical problems that one may loosely associate with philology and editorial practice obtain here; we have no reasonable grounds for doubting that L was written (in a very trivial sense) by A—in other words, A is the author of this document in the minimal sense of having composed this letter and deliberately mailed it to B. Moreover, we know that we have the text of the letter as written by A (i.e., we have the original; we know that we have the integral text; A's hand is fully legible; A wrote no first draft; B in no way altered the document before passing it along to us; and so on). Truly, theory proceeds by means of idealizations!

In relation to this model it is possible to isolate any number of different orientations and questions. If we allow that this model provides a rough and idealized map of some pertinent features of a fictive example of an interaction and exchange by correspondence, we may further specify that other kinds of maps can be situated in relation to it. I want to use this model to give a sense of the selections and exclusions implicit in a certain literary-critical relation to its domain. If we employ this model to ask about the nature of the typical literary questions discussed above, the result may be quite interesting, it being clear that it will be crucial to take note of the many differences between literary practices and an ordinary correspondence. To repeat, I am not setting up this comparison in order to equate all literary phenomena to the type of situation described in the model.

Let us now imagine that an observer-critic undertakes a complex and subtle interpretation of the text of the letter and sets forth the results as a restatement of its meaning or significance. Let us suppose that the critic "discovers" something in this letter, namely, an ironic statement to the effect that B does not really understand *x*. A is thought to be suggesting that B has not really grasped their friend's situation well, or is overly indifferent, and ought to pay more attention to the friend's problem, for example. For the sake of those readers who are loath to entertain what they would call the intentional fallacy, we can imagine that this "irony" in the letter was unintended by A, or that our argument for its presence in the letter does not require reference to an extratextual knowledge of its author (if we need a voice, it could be that of the implicit author-narrator of the letter). Let us also suppose that nothing in the world indicates that B perceives this irony at all, perhaps because of B's perception of A, B's sense of A's viewpoints, B's naïveté, or indifference. In any case, B does not catch the irony, and our imaginary observer has compelling evidence for this fact. Our critical observer, then, finds this meaning in the letter (or in A's intent), can point to evidence for it, but has every reason to conclude that this aspect of its message has no impact whatsoever on the ongoing interaction between the friends; it figures in no good account of what the message is, effectively, for the recipient of the letter. Only the critic-recipient (and perhaps A) think this is part of the meaning of L.

I have constructed this example in order to point to what is left out of the vague question of the meaning of a text's message. To stop with this question is never to argue that what the message has to say is either true or false; nor is it a question of considering whether the letter's putative message makes the least bit of difference to the history of the relationship. This question frequently amounts to an orientation that snatches an utterance out of the history of the interaction in which it is embedded. The message created in this manner may then be amplified and broadcast, but it is seldom made clear that this message is won at the cost of ignoring truth or falsehood and context. This claim can be made more palpable by comparing the situation just described to another one. Imagine that B *does* understand the ironic remark to the effect that B is unaware, indifferent. B takes offense and sends an angry response to A, or tells the friend, who is angered at A as a result—one can imagine all sorts of little dramas being generated by A's ironic insult. And any detailed account of the interaction in these dramas that did not include B's perception of an ironic remark in A's letter would miss an important piece. Still another variation can be used to underscore this point. Let us return to the situation where the irony, perceived by the critical observer, passes unnoticed by B. Let us suppose that we know for certain that A intends this irony and even hopes it is noticed by B. When B does not respond in anger and never makes reference to the insult, A has to interpret B's reaction on this score. A wonders whether B perceived the irony but decided to let it pass without comment. A also conceives of the possibility that B simply did not catch his intentions. Because this latter interpretation best matches A's perception of B, and because the irony is subtle, A adopts this assumption and never again makes any such insinuation. The ironic message, then, makes no difference to their relationship or interaction unless we construe these terms in a very special way. In the example, the irony is definitely part of our observer's experience of the relationship—that point is indubitable. The irony is also part of A's perception of the relationship, but it does not significantly change A's view of B. It is hard to see how we can get this "message" or "meaning" into the history *of the relationship* between two persons when it goes wholly unnoticed by one of them. In such a context of interpretation, this irony is a dead letter.[1] But those literary critics who pose the typical question of a literary text's meaning (Q in the previous chapter) tend to ignore this point. In this manner the door is opened to the most fantastic interpretations of the meanings that texts have supposedly had in history.[2] The critical reader guided by Q feels warranted to engage in the most detailed interpretations of a work's meaning and never checks this wealth of detail in terms of the consequences that the *utterance* actually has within the context of interactions. It is in precisely this sense that the issue of authorial versus textual meaning is irrelevant to the distinction I am drawing here, for in both cases the omission being pointed to is the same. Be the message that of the author or of the text, it is not compared to the effective function or consequences that the utterance or symbolic artifact has within a context of *interaction*. We can at this stage usefully contrast two broad kinds of questions:

Q1: What does a particular text mean?
Q2: What are the consequences of an utterance in the context of the interaction in which it occurs?

What are the differences between these two kinds of question? What is important or interesting about the latter in the case of a literary work? One can easily imagine someone trying to understand

[1]An objection that may be made here concerns the possibility of someone being influenced by a message of which they are unconscious. Music of which I am at most dimly aware may influence my moods; subtle inflections and corporeal expressions influence our understanding of other persons, yet we are hardly aware of them in some fully lucid and or highly conscious manner; messages that were repressed at one moment return in disguised form in a dream. Indeed, all such cases point to complexities that must be explored. But on this issue, one point: the message unconsciously received must still make a difference if it is to figure within any realist explanation of interactions; it cannot be merely the product of the observer's speculative interpretation, and hence its postulation must obey norms of evidence and inference frequently abandoned in psychoanalysis. That the subject agrees with the interpretations is not, in my view, an acceptable proof or confirmation. This is not the place for an extended bout with psychoanalysis; the outcome is in any case predictable. Readers who want a hint are directed to the arguments of Cornelius Castoriadis in "Epilégomènes à une théorie de l'âme que l'on a pu présenter comme science," in *Les Carrefours du labyrinthe* (Paris: Seuil, 1978), 29–64; Mikkel Borch-Jacobsen, *Le sujet freudien* (Paris: Flammarion, 1982).

[2]As an example of the approach I am criticizing here, I would point to Althusser's influential reading of a performance of Brecht in which he claims that the formal innovations in this performance realized a powerful criticism of capitalism, a criticism that possessed this power even though it went totally unnoticed by the public; see Louis Althusser, "Le 'Piccolo' Bertolazzi et Brecht (Notes sur un théâtre matérialiste)," in *Pour Marx* (Paris: Maspéro, 1965), 129–52.

the shift of orientation involved in Q2 by assuming that it is a matter of focusing upon the recipient's meaning or "reader's response." But that is an error, and crucial differences are missed as a result of such an assimilation.[3] To make good on this claim I must give a more precise sense of what the difference is between the kind of question isolated by Q2 and the general question of "response." I formulate this latter question in a very broad way as Q3 so as to give a first approximation of the difference:

Q3: What does an utterance mean to its recipient?

If an utterance can be said to have an effective function or a set of consequences within a particular system of interaction, is this not a matter of the meaning that it has for its recipients? In that case, the last two questions, Q2 and Q3, are essentially the same, except that the former focuses on the first recipients, whereas the latter is a more open and general inquiry into responses to the utterance or text. But there are two separate issues here: the first concerns the difference between an utterance's consequences and its received message in a broad sense; the second has to do with the supposed priority of the original context.

That consequences within a system of interaction cannot be subsumed beneath reception can be illustrated by returning to our model. Let us suppose, as before, that A's letter contains an intended, ironic statement to the effect that B is not really aware of the impact that unemployment is having on their friend. The implication is that B is insensitive, self-centered, and so on. B does not notice the irony—that part of the message simply does not figure in her understanding of it. Thus B's next response to A, and B's ongoing dealings with the unemployed friend, go on wholly unchanged. A takes note of this fact and draws conclusions about B: this person really is obtuse, indifferent, and so on. As a result of these conclusions, A has a different attitude toward B, and their relation changes drastically; for example, A might reveal all of this to the friend and could end up persuading this person that B is no friend at all. Should we now try to analyze the history of the interaction, the ironic message suddenly has consequences, for as a message that should have been noticed, it helped to bring a difference within the relationship. It makes a dif-

[3]Here I am, of course, brushing against the large and complex literature of reception criticism; see Robert Holub, *Reception Theory: A Critical Introduction* (London: Methuen, 1984).

ference to the history of the interaction, and there is a change in its basic patterns. This difference is not seized if we simply turn our attention to the reader's response. This difference is not to be found *in* B's understanding of the letter. In fact what gives rise to A's change in attitude is a consequence of B's actions that is wholly unintended by B. B has not got a clue about the irony and therefore cannot understand the impact that her next letter has on A. But B's response has everything to do with the fact that in this variation on the example the unnoticed irony makes a difference. Or imagine yet another variation. B in fact notices the irony but decides to behave as if she has not noticed it at all. A, however, reads this behavior in the same way as just described, that is, as definitive proof of B's self-centered and obtuse character. Thus the relationship changes exactly as described before. Once more, even the most perfect interpretation of the letter exclusively in terms of either A's intended meanings or B's reading does not lead to an adequate explanation of the history of the interaction, for it is not the irony that makes the difference. Nor is the history of the interaction, the course of the relationship, decided by some simple combination of what A thought and what B thought, of what A intended and what B intended. The intended results of an action, without which it is meaningless to speak of purposive action at all, must be distinguished from an action's various consequences.[4] We should note the role played by unintended consequences within an interaction, effects that, although intended by neither party, amount to changes in their ongoing interaction and relations.

But what kind of difference is at stake? It should be clear by now that the difference I am trying to isolate, a difference not singled out by questions Q1 and Q3, can only be plotted in relation to a body of theoretical and factual knowledge that gives some precision to the notions "context" and "interaction" figuring above. The difference at stake here stands out against a prior conception of which agents are involved in a system of interaction, as well as other information about the setting, history, and detailed circumstances of that system. Moreover, the difference in question requires an understanding of what it means for a relationship between human beings to change or remain the same. The notion of consequences employed in Q2 would have to be filled in in such terms, the basic point being that the

[4] G. H. von Wright, *Explanation and Understanding* (Ithaca: Cornell University Press, 1971), 67; see also Raimo Tuomela, *A Theory of Social Action* (Dordrecht: Reidel, 1984), chap. 3.

notion of the meaning or message of a text is wholly general, whereas that of an utterance's effective role in the history of an interaction is a special case defined in terms of conceptions of specific agents, histories, interactions, and contexts. The former is an indeterminate question, whereas the latter is determinate relative to the specifications of context, agents, and so forth—which is not to say that the latter question can always be given a reliable answer.

I must stress that an utterance's consequences are not simply the "sum total of what happens because of the utterance." Such an interpretation leads to the wildest absurdities, and it is easy to think of examples of the misapplication of such a topic. Imagine that B receives A's letter, takes note of the irony, feels justly castigated for her indifference to the problem of unemployment, and goes into a state of depression. As a result, she fails to go to work one day and is replaced on the job by someone else, who had been planning to do something else. Would a legitimate reading of the letter's consequences within the history of the interactions in which it first appears include all of these events and those to which they in turn give rise? Not at all. What must be noted is simply the necessity of far more careful and realistic delimitations of the interactional system under investigation. In the previous example, one could decide to focus on the interactions between A and B and thus exclude from consideration the other consequences. In the context of an investigation of the history of A and B's interactions, it is clear enough that the letter alone is not *the* cause of B's depression. The key, then, is the delimitation of the systems of interaction pertinent to given inquiries, for in its absence history can only be of the Shandean variety.

Literature in Context, Criticism as Context

It is possible to distinguish between an analysis that centers upon the general or received meanings of a text and an inquiry into its consequences within different, specifically identified interactional systems. But the possibility of such a distinction tells us nothing about the relative merits of this and other interpretive orientations, nor does it necessarily have any great pertinence for literary criticism. Given these considerations, it is necessary to turn directly to the supposed priority of the original or first context in which an utterance occurs. Why would a system of interaction defined along these lines be more important to critical analysis than any other set of interactions in which a text may figure? Cannot the meaning of a literary text have an effective function or important consequences in the relationship between various living critics and a dead author, who addressed his or her work to an unspecified audience? Are not *we* ultimately responsible for any arguments about the validity of this or that interpretation? The critic who thinks, for example, that the correct understanding of *Hamlet* is that of the Elizabethan world picture owes us not only a reading of this latter but an argument why this "original" moment of reception should be privileged. Moreover, the idea of privileging an original context, the one in which the utterance was first made, seems to be particularly inappropriate to the example of a *literary* communication. The case of the correspondence between friends in that instance serves most of all to bring forth the crucial differences between literary and other contexts of interpretation. Whereas in the one case we have a letter addressed to a specific individual who is clearly its first recipient, a recipient whose relation to the sender is easily isolated as a pertinent unit of analysis, in the other case we have an open-ended form of communication. Imagine insisting that one must read Franz Kafka's *Brief an den Vater* only in terms of what it can tell us about the history of the Kafka family. That would be even more absurd and inappropriate than using the author's writings as materials in a clinical diagnosis of his individual "case." Surely our interest in literature ought to lie elsewhere.

John Ellis, for one, seems to make this difference the crucial defining characteristic of the literary *approach* to a text. Thus, he defines a literary text as one that is not taken as being specifically relevant to the immediate context of its origin.[5] In this case, the literary text and message lacks any specific addressee, or better, even if it has a specific addressee in the writer's immediate public, it cannot in principle have *only* this public, for its significance and value transcend what Ellis calls the immediate context of its origin. Whether an author claims to be writing *für Ewigkeit*, as Gramsci put it, can never settle this matter, for what is under analysis is a social and historical process of literary meaning, and only the end of this process, the end of history, can seal the envelope of reception once and for all. Even more important, what is in question is an *attitude* taken toward the text which, if it could be studied in its connection to an original context, would cease in that moment to be perceived in terms of its literariness. Thus if Q2, or some rigorous version of it, designates a

[5]John Ellis, *The Theory of Literary Criticism* (Berkeley: University of California Press, 1973), 112.

valid type of inquiry, it should be recognized that this is *not* the kind of research appropriate to the literary disciplines.

At this stage in the discussion, the question of the critic's interrogative priorities is directly posed, for it appears that the choice of levels and units of analysis in matters interpretive has to do, first of all, with a choice concerning which groups and agents, which institutions and sites of interaction, are to be placed in the foreground. One can indeed speak of an analysis that seeks to discern the effective function of a particular literary work in the context, not of its original utterance, but of an interactive system defined as that of the history of literary criticism in relation to this text. Presumably all those who have written in response to this work, proffering commentary, judgments, and so on, constitute the group under consideration, and the text's effective function is its impact upon this group—not only the meanings they ascribed to it, but whatever other changes and differences might be the consequences of the work's reception. Yet this example already seems terribly artificial—its assumptions about the history of criticism and its institutions are inadequate. To speak of an analysis that orients itself toward the historical consequences of an utterance is to think in wholly different terms, for though such an analysis still deals with the meanings and the messages of a work or text, these are taken as means toward another end, that of discerning the history of the interactions between human individuals and groups, interactions involving institutional settings and norms. One way to put this difference, hastily, is simply to say that it is a dubious sociological and historical premise to assume that a system of literary interactions exists in autonomy from the other institutions, norms, and sites within which the members of the same society are engaged. Art cannot exist independently of the collectivity, and the forms of its dependence are several.[6] In this regard, I cannot fully share the opinion of Siegfried Schmidt, although I agree with him on the crucial point that literary history should not be a history of autonomous texts and must be construed as "histories of text-oriented actions and their full scale set of action conditions."[7] It strikes me as misleading to hold that the literary interactions form in any real sense a unified or autonomous *system* and therefore constitute a separate domain to be investigated. It seems wrong to pro-

[6]H. S. Becker, "Art as Collective Action," *American Sociological Review*, 39 (1974), 767–76.

[7]Siegfried J. Schmidt, "On Writing Histories of Literature: Some Remarks from a Constructivist Point of View," *Poetics*, 14 (1985), 279–301, citation: 292–93.

claim that "the art system in Europe, e.g., has been freed from religious and political interests since the 17th century."[8]

One form of analysis singles out certain texts as autonomous literary works to be interpreted in relation to a community of literary critics. The other orientation seeks to isolate other systems of interaction in which texts and utterances are engaged. What are the relations between the two orientations? This question is extremely difficult. It may be first approached by pointing out that some fairly dubious assumptions would be required for us to consider that the two explanatory contexts or interpretive orientations are ultimately the same, or that the choice between them is arbitrary. The difference between them can be brought forth even more sharply if we entertain another type of question, wholly complementary to Q2:

> Q4: What systems of interaction are the proximate conditions of the emergence of the utterance?

This question focuses research on the conditions of the genesis or emergence of texts or utterances. It is complementary to Q2, our investigation into consequences, for in both cases we isolate the same interlocking network of institutions, norms, and relations—institutions and relations that are not likely to be purely literary. The alternative orientation involved in both Q2 and Q4 brings with it wholly different criteria concerning the validity of an interpretation, as well as rather different expectations about the pertinence of the results. To arrive at even an approximately accurate understanding of the place of a work within concrete interactional systems requires a reliable knowledge of the social history in question. Thus such an interpretive endeavor must be guided by the best available knowledge of this social landscape, for it must draw upon theories of social interaction as well as upon historical theses about the nature of the particular circumstances, institutions, norms, and agents in question. A frequent error is the assumption that this sort of approach to the interpretation of literary works amounts to Marxist aesthetics, and once this equation is made, the rejection of the reductive sociologism or even economism whereby the cultural domain is supposedly explained by reference to some dubious laws of history. Although it is true that some contributions to the Marxist tradition

[8]Schmidt, "Writing Histories," 293. It should be pointed out that Schmidt in general advocates a properly historical understanding of aesthetic conventions; all I claim is that a more thorough application of his own perspective would contradict his broad assertion of the autonomy of the art system.

are reductive in this manner, subtle and up-to-date work in the Marxist vein reveals the emptiness of such accusations. Moreover, the basic orientation being proposed here does not require the adoption of any particular theses about the nature of society, history, or culture.[9] It *does* require a sociology and a history of some sort, which is why doing this kind of work is so forbiddingly difficult, particularly in relation to anterior historical periods about which our knowledge is all too sketchy. But there is probably another reason why such attempts are infrequent in literary circles, for it is evident that in relation to such an interpretive strategy, the literary is radically decentered and must clearly sacrifice any pretension to disciplinary autonomy. I cannot possibly understand the complex history of the conditions and consequences of what we glibly label *Le Tartuffe*, in France from 1664 to 1670, without relying extensively upon any number of nonliterary hypotheses and documents. But this is only a first decentering of the literary discipline, and there is yet another one waiting in the wings once we begin to take this kind of alternative orientation seriously. It will frequently turn out that the details of texts are not always so very decisive in questions about the history of social relations and interactions, of conflicts and domination, of institutional stability and changes.[10] Already in the schematic examples analyzed above, cases were advanced where the presence of an ironic message in a text—even when solidly verified from a document-oriented perspective—was simply not a decisive factor in an interactional history. Literary critics who begin to take an interest in an alternative approach to history, that is, one that does not give

[9]What strikes me as indispensable in the Marxist tradition of thought is its insistence upon a realist epistemology, its materialism (construed in a nonreductionist manner), as well as an initial identification of certain key problems. Yet these real virtues of the Marxist sociological tradition do not make it a self-sufficient program of explanation in the sociohistorical domain. I am particularly persuaded, for example, by Cornelius Castoriadis's radical questioning of certain Marxist theses about the nature of history and the theory of value; see his *L'Institution imaginaire de la société* (Paris: Seuil, 1975), 13–156; as well as his "Valeur, Egalité, Justice, Politique: De Marx à Aristote et d'Aristote à nous," in *Les Carrefours du labyrinthe*, 249–316. Some of the most vital issues in social theory are sharply delineated by Jean-Pierre Dupuy in his "L'Autonomie du social: De la contribution de la pensée systématique à la théorie de la société," *Cahiers du C.R.E.A.*, 10 (1986), 229–73.

[10]That the role of text-oriented analyses within a sociological approach to literature is not obvious has been noted by Claus-Michael Ort, "Problems of Interdisciplinary Theory-Formation in the Social History of Literature," *Poetics*, 14 (1985), 321–44, esp. 331. On this score, any pretensions about deriving the consequences of texts from stylistic analyses should be contrasted to the suggestive contentions of Vladimir J. Konečni, "Elusive Effects of Artists' 'Messages,'" in *Cognitive Processes in the Perception of Art*, ed. W. Ray Crozier and Anthony J. Chapman (Amsterdam: North Holland, 1984), 71–93.

itself in advance the task of snatching works and messages out of networks of action, may find that the literary a priori does not serve as a reliable guide to those texts and events that have made significant differences in the history of societies and of interactions.

I have been arguing for the possibility of a rather different approach to the interpretation of cultural artifacts, one that seeks to read them in terms of their roles within the context of their emergence. I am not arguing that this is the only valid approach to these artifacts—such an idea would be a gross misreading of what is being said here. As I have pointed out repeatedly in this work, scientific research finds only a first and broad directive in its goal of obtaining true and systematic knowledge; other priorities and interests must orient specific investigations. Only such interests could motivate a critic to adopt the kind of approach that I have identified in Q2 and Q4, just as a certain kind of aesthetic a priori typically motivates Q1. What I would like to challenge, however, is the idea that there exists a set of objects that can only rightly be approached by means of the aesthetics discussed above. In this regard, it is crucial to note that the simple fact that a work does not have a true, *literal* reference hardly indicates that it has no conditions or consequences within social interaction. It is a basic error to think that literary utterances are "pseudostatements" having no literal truth function, somehow sealed off from the pragmatic context in which they occur. Nor is it enough to recognize such a context but to go on to identify it as one where no immediate utilitarian purposes, such as exhorting action and conveying information, are involved. In Molière's context "Tartuffe" was not a name that referred literally to a single person, but this fact hardly settles the question of the play's reference, which involves *abstract* and *typical* situations, individuals, and events.[11] It is also wrong to speak of a breakdown of the proper aesthetic distance in relation to the events surrounding this work. Are we to castigate the churchmen who condemned the play because they did not focus properly on its literariness? Did they foolishly mistake fact and fiction, the literal and the metaphorical? But the same criticisms would have to be applied to Molière and his allies as well, for although they may have been interested in using the framework of a

[11]That the literary-nonliterary distinction cannot be correctly based on literal truth values is argued by Mary Louise Pratt, *Toward a Speech Act Theory of Literary Discourse* (Bloomington: Indiana University Press, 1977), 143. The background to my semantic distinctions is Jon Barwise and John Perry, *Situations and Attitudes* (Cambridge, Mass.: M.I.T. Press, 1983); "Shifting Situations and Shaken Attitudes: An Interview with Barwise and Perry," *Linguistics and Philosophy*, 8 (1985), 105–61.

theatrical *fiction* to make mockery of certain types of religious attitudes, the play would probably not have been of great interest to them had it been perceived as a pure fiction bearing no reference whatsover to the realities of the time. And we must note how little the pragmatic consequences of this reference can be arrived at through an analysis of only what was literally depicted on stage; even after "Tartuffe" had been renamed "Panulfe" and the play significantly reworked, the new play *L'Imposteur* was still banned by the Parlement of Paris and censored by the Archbishop of Paris, who threatened with excommuncation anyone who listened to or read the work.[12] The moral of my story is that the pragmatic role of an utterance or text is not something that can be read out in a purely semantic analysis, and that the latter sort of approach can be considered sufficient only within an orientation that divorces itself from historical realities. What is more, the aestheticized approaches to literary semantics are not even adequate to the question of the work's reference in any absolute sense, because the question of the truth or falsehood of a given message is not even seriously posed. To determine whether any of the versions of *Le Tartuffe* accurately refer to the typical foibles of a certain type of religious devotee (the so-called *petit collet*) or the typical relations between the believers and the *directeur de conscience* requires moving outside these texts to study others illuminating aspects of seventeenth-century French society.

Literary Knowledge

As I indicated earlier, it is my intention here to single out two very different avenues of literary research which I feel deserve greater attention. The first concerns the integration of interpretation within the contexts isolated by sociohistorical research. The second has to do with what I perceive as being the epistemological value of readings of literary texts. As I have contended several times, one of the consequences of a certain aesthetic a priori of critical interpretation is the putting in brackets of the work's message: its sense may be elucidated at great length, but it is assumed that fictions have no

[12]For background, see Georges Couton, "Notice" to *Le Tartuffe*, in Molière, *Oeuvres complètes* (Paris: Pléiade, 1971), 1:833–81; Herman Prins Saloman, *Tartuffe devant l'opinion française* (Paris: Presses Universitaires de France, 1962); Gustave Michaut, *Les Luttes de Molière* (Paris: Hachette, 1925); and Karl Mantzius, *Molière and His Times* (Philadelphia: Lippincott, 1906).

literal reference and hence should not be examined in such terms. The work's or artist's vision may be thought utopic, morally relevant, creative, original, and so on, but it is rarely argued that it is *true* (although many critical works seem to rely implicitly on some such claim). But what would making any such additional judgments and arguments entail? It is clear that, in order to say something about what is the case, it is not sufficient to explicate some remarkable individual's vision, for that vision must be held up against other accounts and, ideally, against the states of affairs to which these accounts refer. But at that moment the literary text is decentered, it is not approached "in itself and for its own sake," its indubitable aesthetic perfection as an end in itself is replaced for its rather uncertain value as a means to something else. This seems a rather precarious way to defend the importance of literature, for as Hegel well noted, once we start to evaluate something as a means to an end, we can only give it our rational preference if it is the very best (most efficacious) means to that end. In the case of art, it seems odd to imagine that a mere appearance (*Schein*) should be the best means to the end of serious knowledge. Have not thought and reflection indeed "spread their wings above fine art?"[13]

Must the artist's vision be tested against other discourses, then? No such evaluation is the least bit *necessary*, and indeed I would be the first to advise any critic whose primary interest is the *art* of literature to eschew such matters. But the critic who wishes to make claims about literature as a form of knowledge cannot afford to avoid this challenge. It is the critic's research into the meanings of literary visions that must be tested against other discourses. Moreover, this challenge is leveled particularly at the critic who already goes to great lengths to elaborate upon the meanings and messages of literary works, to the point of disfiguring their aesthetic features and of ignoring their historical conditions and consequences. I also insist that to speak of "testing" the literary messages against other discourses is not a matter of proposing that some extraliterary doctrine or discipline be established as *the* authority in relation to which critical interpretive knowledge must be checked. The model of knowledge implicit in the natural sciences, and elucidated in various ways by philosophers, is an excellent debating partner for critical knowledge claims, as are different research programs within the social sciences, psychology, anthropology, and linguistics. One

[13]G. W. F. Hegel, *Aesthetics: Lectures on Fine Art*, trans. T. M. Knox (Oxford: Clarendon, 1975), 1:4–10, 51.

should not, in any case, assume in advance that the cognitive exchange between literary criticism and these research programs will move in only one direction. This sort of one-sided influence, which is at once wrongheaded and useless, is what happens when the literary critic, having assimilated the current doctrine of some theoretical trend within one of these areas, proceeds to *apply* these "truths" in a reading of literary messages. Such a procedure amounts to a betrayal of the cognitive potentials of criticism, literature, and the doctrines copied. What I have in mind is rather different, for it involves an investigation of the ways oriented readings of literary works serve to challenge and to refine, to complexify and perfect hypotheses within the other anthropological disciplines. This type of inquiry is rather different from any of the questions listed above and, when stated in a formula, amounts to something like the following:

> Q5: What is the meaning of a text in the context of a particular program of research (typically, a body of hypotheses and evidence within one or more of the anthropological disciplines)? In what ways can the interpretation of the text contribute to the research program through a refinement of its hypotheses?

A few remarks on this formulation are needed. The emphasis on the refinement of hypotheses is made for two reasons, having to do first with the nature of the contribution that critical interpretations of literature are most capable of making, and second with the kind of contribution that is typically most badly needed in the anthropological disciplines. The idea of a refinement, complexification, and challenging of existing hypotheses covers both. It should be evident that the critical examination of literary works cannot be viewed as a possible contribution to empirical research, except within the context of the kind of historical inquiry associated with Q2 and Q4. In this latter context, Molière's *Le Tartuffe* and related documents are viewed as facts within the historical domain being explored, and interpretive inquiries into them are as empirical as anything else ("empirical" here, of course, has nothing to do with any phenomenalist biases about the five senses, or with myths of direct observation). But in the context of an inquiry falling beneath the rubric of Q5, the status of a literary work is wholly different. Here it has no major effective function or important consequences within a factual domain, and the consequences it may have are simply not pertinent. Rather, the role of textual analysis is in this context heuristic, epistemological, and cognitive.

An example may help to make the difference clear. Let us imagine a broad research program in the sociohistorical sciences having to do with structures of kinship and exchange, with a particular historical focus on the emergence and nature of the Western institution of the family. Let us further imagine that one of the foci of this research concerns the relationship between kinship structures and relations of exchange in a more general sense (including gifts, rituals of potlatch, barter, trading, market relations). In its approach to the modern institutional forms of marriage, then, this line of inquiry is interested in exploring, for example, those aspects of matrimony having both an explicit and an implicit economic dimension—an obvious instance being the practice of the dowry. Another point to be interrogated could concern the respective statuses and roles of matrimonial partners in relation to the external marketplace, the asymmetrical dynamics generated within a tradition where the wife does not engage in productive or profitable labor, for example. I cannot imagine how any literary text, or corpus of literary texts, could ever serve as a telling *example* or decisive body of evidence in relation to such matters. The cleverest readings in the world of any number of novels and plays dealing with such topics would never establish anything, for the orientation of the research is not primarily toward how the society in question referred to itself (in fiction or elsewhere) but toward the effective determinations of social life. This social inquiry wants to know many things: it may want to discover what typical patterns of interaction are constituted by a particular set of norms and institutions; what kinds of creative resistence and innovations individuals adopt in relation to those institutions; what the consequences for human life are—what potentialities were developed, and which were repressed. The literature of the period may have had some real consequences in this regard, but it would be dubious sociology indeed to assume that the institutions of art were somehow determinant or decisive. Literature and the arts are not excluded, but neither are they shoved into the foreground in some arbitrary and a priori manner. But I have said all of this, not in order to affirm that literature and its study can have no role in relation to such inquiries, but merely in order to sharpen the conception of what this role may be. What I want to stress, then, is the *epistemic* nature of this contribution made by a literary knowledge. The literary work is not viewed in such a context as an example of a thesis, nor even as a more sensuous, detailed, or particularized illustration of social realities. Rather, it is approached as the manifestation of a useful and revealing *perspective on* the problem with which this research is concerned. In regard to the example sketched above, a reading of Emile Zola's *Pot-Bouille* in relation to the body of social hypotheses could seek to discover the ways that text's implicit model of the matrimonial economy (and its relation to realities economic *stricto sensu*) amounts to a valuable commentary on the same issues, a commentary that could in fact lead to a complexification of the hypotheses and theoretical assumptions necessary to any real social inquiry. Whether this commentary and model are those of the departed empirical subject Emile Zola is anything but central to the kind of research I am delineating. The critic who follows my suggestion should have a great deal of freedom in the choice of materials, so that if the notes, documents, letters, or other works in the corpus reveal anything of pertinence to the reconstruction of the model, they should certainly be added. But to demand substantiation of the commentary with evidence suggestive of the author's intention is merely to place an arbitrary constraint on this kind of inquiry, a constraint that may rightly be applied to define another context of research. The difference is a question of emphasis: what matters in one context is what *Zola* believed, and in the other, what valuable *beliefs* made be reconstructed through readings of his (and other) writings.

This type of contribution to the human sciences not only matches one of the potentialities proper to the critical analysis of large number of literary works, but it also responds to an objective problem within the human sciences. In this regard, literary knowledge is not the supplicant seeking a little corner for itself within the marvelous edifice of the social and human sciences. On the contrary, the sorry state of the hypotheses and theoretical models guiding many of the specific research efforts within the latter recommend the importance of broadening the general consensus about the cognitive resources available to those interested in social and historical inquiry. Here I return to Castoriadis's remark about the poverty of the psychologies that underwrite the models of orthodox economics, a poverty that is notably absent in the various models of desire, exchange, interaction, power, and gender relations implicit within a large part of literature. The case of economics is in no way special, and many other areas could also be pointed out.[14] My hypothesis, then, is that literary texts can be very usefully mobilized in relation to the models and assumptions at work in a range of anthropological research pro-

[14]An example is the research in artificial intelligence and cognitive science on discourse processing and story comprehension. Literary critics could learn from, and contribute to, this research program. See my "Théorie littéraire et sciences cognitives," in *Approches de la cognition*, ed. Daniel Andler (Paris: Seuil, forthcoming).

grams, particularly in regard to social psychology, sociological and philosophical theories of action, all of the topics of sociological theory, political philosophy, law, individual psychology, and ethics. I do not think, however, that literary works are likely to illuminate the issues of quantum mechanics, nor that the problems of the latter are particularly revealing in regard to the epistemology of the human sciences. But this is hardly a telling restriction, and it leaves the literary researcher interested in this particular line of inquiry with a lot of interesting encounters to make.

One point implicit in this discussion should perhaps be avowed quite openly. The kind of contribution being discussed here amounts to an *immanent* critique of the hypotheses and models employed in the extraliterary research programs, by which I mean to say, not that the role of literary knowledge is purely negative, but that its criticisms and amendments arise in a determinate relation to the existing research strategy and hence are oriented toward its particular anomalies. If the type of research I am advocating is to be a viable institutional strategy, it cannot be exemplified by the maverick literary scholar who posits a new economics, linguistics, or psychology without engaging in an extensive and detailed dialogue with the state of the art in the relevant fields. Nor can it be an interdisciplinarity that flies in the face of the specific problems and evidence that have required a specialized and particularized attention. The literary text and its interpretation must not be used as a playground within which the critic's imagination toys with the instruments of science. To treat it as such is to betray literature's potential as a productive and creative space where pertinent hypotheses and models are developed—and where the critic may hope to make effective interventions in a knowledge of man and society which is, at the present time, far too inadequate.

Priorities: From Pure Aesthetics to Literary Knowledge

As I have frequently noted in this book, the pursuit of knowledge is necessarily oriented toward finding out the truth, but this orientation is too loose and general to provide any more specific guidelines concerning the actual direction of our inquiries. Thus the problem of a discipline's epistemological priorities cannot be avoided; it is a problem that is dealt with either in a blind and implicit manner as a result of a series of "invisible" theories, or consciously and as a result of explicit discussions and choices. In this final section of the work, I try to make explicit my own sense of the interest of some of the different possibilities that presently stand before the literary disciplines.

It should be clear by now that I think that the task of producing aesthetically oriented interpretations or readings has had an undeserved priority within modern critical practice. To state my objection to this type of work as briefly as possible, I would say that it falls between stools, for it is at once an aesthetics, a social history, a biography, and a model of communications, but it combines elements of all of these orientations in an odd, and not very successful, mixture. In its rather industrious and pseudoscientific efforts to discover the message in the bottle of literary art, it has betrayed the priorities of one kind of aesthetic attitude toward literature, which has little to do with a series of interpretive raids upon the works' messages—messages that are, by the way, "with more spirit chased than enjoyed." Yet because the business of interpretation has at the same time held fast to some of the assumptions of a certain aesthetics, the meanings it discovers are held frozen within the brackets of fiction, and literary knowledge remains a diminutive craft, a delicate ship-in-the-bottle, in which no one would think to set sail on the seas of reality. The carving out of an aestheticized framework of interpretive practice has had a dual effect: first, such research has frequently been cut off from the possibility of fruitful encounters with the other anthropological disciplines—the result being a loss to all parties; second, it has tended to make the literary artifact look like something existing within its own autonomous sphere, a sphere where its varied meanings and patterns may connect and disconnect following myriad pathways, but where none of these semantic choices constitutes an effective difference. Thus the cognitive possibilities of literature, possibilities that extend across contexts utterly remote from that of the author and an initial audience, are restricted to *one* type of context, to one putatively self-sufficient and autonomous institution, that of the professors of literature. But in regard to the question of what constitutes knowledge, it is pure folly to believe in that discipline's autonomy or in its self-sufficiency. The history of criticism in the twentieth century, which is the history of this kind of professionalized interpretive practice, is a history of the borrowings, both creative and sterile, whereby the study of literature has been oriented and stimulated by a range of ideas, theories, and orientations arising from within other fields. A specifically literary reading of a poem, novel, or play simply does not exist, once we subject this common notion to rigorous examination—and the history of efforts to base a literary inquiry upon literariness reveals that the searches for the proper essence of the literary were always underwritten by some combination of linguistics and philosophical aesthetics: first a theory of the aesthetic function, and second a focus on the dialectics of linguistic devices in ordinary and poetic spheres.[15]

These criticisms are not meant to deny the possibility, interest, and value of an aesthetics of literature, that is, one that to a certain degree puts in brackets the utilities and validities of the literary work and concerns itself instead with matters related to aesthetic value. But that sort of work, which must not be confounded with the elaboration of detailed and far-reaching interpretations of a work's implicit meanings, is not in fact what most critics seem to be doing today. Perhaps the literary critics who contribute their efforts to the hermeneutic mills of the meaning-making industry in fact have aesthetic experiences of the literary works before and after they write about them, but the relationship between their scholarly writings and the aesthetic experience is not obvious. The knee-jerk answer that doing detailed interpretations enhances the next moment of aesthetic appreciation may be plausible in relation to a certain level of pedagogy, but it would require a lot of work to be applicable to the fruits of advanced interpretive research. Rather, it seems to me that the critic's own prose is more and more set forth as an end in itself, the motivations behind this being largely pragmatic. But in this way, neither knowledge nor literature is served. What requires a closer examination, then, is the legitimacy, pertinence, and value of an interpretive orientation resulting from an uneasy combination of aesthetic and sociohistorical perspectives, for my hunch is that this combination may not be advantageous for either of these two interests. But of course one may argue that the megaphone response, this relating to the artwork as an oracle in brackets, is what has become of aesthetics, a concept that is an essentially historical and relative label for what the producers and consumers of art typically do in any given context or period. Yet what may be noted as we seek to characterize this new aesthetics is its strange insistence upon the importance of finding, or failing to find, elaborate meanings in works of art. Should an aesthetic approach to literature lend priority to notions of meaning and communication at all, and does it really find its culmination in the art of paraphrase? In any case, one of the tasks of critical

[15]An excellent source on one aspect of the history of such research is Frantisek W. Galan, *Historic Structures: The Prague School Project, 1928–46* (London: Croom Helm, 1985).

theory must be to inquire further into the status of the aesthetics that could underwrite arguments for this kind of literary value.[16]

To recapitulate, I have distinguished above between aesthetic interpretive practices and those that situate literary texts within some determinate context in which the history of the social interactions involved are isolated and made the focal point. Although I am by no means wedded to these terms, I have tried to suggest the value and importance of a criticism that delineates the conditions and "consequences" that literary works have had in specific historical contexts, or the "effective function" of literary practices. While I acknowledge that the meaning of a text or utterance is still at stake in such an inquiry, I want to stress the extent to which a text's conditions and consequences cannot be grasped in the absence of a properly sociohistorical knowledge of the agents, institutions, and norms within which the text emerges and where it plays its role. This kind of inquiry is not reducible to the issue of how a work or author draws upon the anterior tradition, nor to the question of what it meant to the history of its reception. Rather, it amounts to the study of literary documents within the contexts of the best available hypotheses and evidence provided by the sociohistorical disciplines.

Second, I want to stress the epistemological value of a literary knowledge that finds another of its multiple specificities in an active encounter with the issues within a range of research programs within the human sciences. This suggestion is not a call for interpretive illustrations and applications of the doctrines made available by a range of superior disciplines. Rather, it is possible for critics to demonstrate the pertinence and unique value of the models, insights, and critiques implicit in literature. What I have in mind is the possibility that literature be employed to complexify, challenge, and improve the models and hypotheses guiding research within the humanities and social sciences. This possibility is not a new idea, but I do believe that it has not been adequately explored. Evoked by Aristotle's passing remark that the potential truth of poetry is more philosophical than historical, the idea was broached in the context of modern science by Beardsley in a passage that deserves to be cited:

> The literary work does not, of course, give us the evidence, and without the evidence the hypothesis can hardly be called knowledge; still, in scientific inquiry the hard thing is often not the testing of a hypothesis

[16]Two valuable sources are Siegfried J. Schmidt, ed., *Literatur und Kunst—Wozu?* (Heidelberg: Carl Winter, 1982); and Stefan Morawski, *Inquiries into the Fundamentals of Aesthetics* (Cambridge, Mass.: M.I.T. Press, 1974).

> once we think of it, but the thinking of an original and fruitful hypothesis in the first place. Therefore literature may have immense cognitive value even if it merely suggests new hypotheses about human nature or society or the world, and even if only a few of these hypotheses turn out to be verifiable, perhaps after some analysis and refinement. One of the claims most often made for literature, that it increases our understanding of human nature, can be thoroughly justified in this way.[17]

But what does the word "literature" refer to, finally? To that question there can be no determinate answer, for it is exactly like asking the physicist to define the total reality of the concrete pendulum. Thus I can do no better than to repeat the sprawling definition already provided by the marvelous and mediocre Jean-François Marmontel, whose entry in the *Encyclopédie* under "littérature" begins: "a general term, which designates knowledge of belles-lettres and related subjects."[18] This is the kind of definition we require today, provided that we recognize the need for creating precise and useful local inquiries within this global discipline that potentially comes into contact with all of our knowledge.

[17]Monroe C. Beardsley, *Aesthetics: Problems in the Philosophy of Criticism* (New York: Harcourt, Brace, 1958), 429–30.

[18]*Encyclopédie ou dictionnaire raisonné des sciences, des arts, et des métiers, Nouvelle impression en facsimilé de la première édition de 1751–1780* (Stuttgart and Bad Cannstatt: Friedrich Frommann, 1966), 9:594–95.

Index

Achinstein, Peter: 64n.19, 225n.31, 226n.33
aerodynamics (aviation): 74–75
Aeschylus: 10
aesthetics (philosophy of art):
  and interpretive criticism: 222, 265–66
  and literariness: 216–18, 244
  Marxist: 255–56
  organicist: 139–41
  romantic: 218–24. *See also* art, literary criticism, literature.
Albert, Hans: 41
Althusser, Louis: 249n.2
Altieri, Charles: 158n.8, 210–11n.9
"always already" argument: 5, 9, 176–86
anomalies: 109, 228–30
Apel, Karl-Otto: 40, 49n.7, 108–9, 113
Ariew, Roger: 91n.22, 102n.38
Aristotle, 20, 55–56, 185, 211–13, 266
Arletty, 51
Aron, Thomas: 214n.18
Aronson, Jerrold L.: 83n.5
arrhepsia: 61
art:
  autonomy of: 219–23, 254–55
  institutional theory of: 223, 254–55
  as *Schein*: 259. *See also* aesthetics.
Ashby, William Ross: 205n.2
aspect-shift: 92–93
Atkinson, R. F.: 231n.38
Attila the Hun: 39
Bacon, Francis: 102
Baker, G. P.: 162–63, 164–65
Bandura, Albert: 171n.24
Baricelli, Jean-Pierre: 19
Barthes, Roland: 29n.27, 203n.1
Bartley, W. W. III: 40n.2
Barwise, Jon, 257n.11
Bateson, Gregory: 207
Bayertz, Kurt: 26n.21
Beardsley, Monroe C.: 223, 266–67
Becker, H. S.: 254n.6
belief:
  and authority: 58
  basic: 42
  distinguished from knowledge: 40–41
  holist networks: 60, 173
Bell, Richard H.: 168n.19
Benjamin, Walter: 28
Berger, Peter L.: 72
Bernstein, Richard J.: 98n.32
Bertalanffy, Ludwig von: 140
Bertilson, H. S.: 115n.6
Beversluis, John: 168n.19
biology: 21, 131–32. *See also* evolution, organicism, self-organizing systems.
Black, Max: 166n.17, 208–9n.7
Blackwell, Richard J.: 80n.1
Blanchot, Maurice: 196
Bloor, David: 67, 70n.28, 71–74, 78n.35, 161–62, 171n.25
Bohm, David: 231n.38
Borch-Jacobsen, Mikkel: 249n.1
Borgeest, Claus: 11n.5

Index

Borges, Jorge Luis: 207
Boulby, Mark: 219n.23
Bourdieu, Pierre: 70n.28
Bouveresse, Jacques: 168n.19
Boyd, Richard N.: 5, 73n.33, 81n.3, 83n.6, 90, 91n.21, 98–102, 103, 104n.43, 106–7, 188n.45
Boyer, Alain: 147n.1
Brahe, Tycho: 93, 100
Brand, Myles: 189n.46
Brecht, Bertold: 249n.2
Brief an den Vater: 253
Bromberger, Sylvain: 227n.34
Bunge, Mario: 73n.33, 88n.16, 102n.37, 105n.44, 125n.3, 209n.8

Caillois, Roger: 160
canon: 6, 13
Carlson, Marvin: 14n.7
Carr, Brian: 36n.2
Castoriadis, Cornelius: 40, 80n.2, 112n.3, 198, 249n.1, 256n.9, 262
causality: 73, 109–10, 227–28
certainty: 164–68. *See also* justification, knowledge.
Chomsky, Noam: 191
Churchland, Paul M.: 86, 89, 135
Cioffi, Frank: 168n.19
communication:
  basic model of: 245–46
  transcendental a priori: 39–40
constitution: 125–27
constructivism: 81, 87, 106–7
  basic tenets of: 91. *See also* framework relativism, instrumentalism, philosophy of science.
context of discovery, context of justification: 69–70, 73, 150
contradiction. *See* logic.
Copernicus, Nicolaus: 78n.35
Couton, Georges: 258n.12
critical theory. *See* literary theory.
Critical theory (Frankfurt school):
  on science: 108–9. *See also* Apel, Karl-Otto, Habermas, Jürgen.
Culler, Jonathan: 11n.5
Cunningham, Frank: 111–12n.1

Darwin, Charles: 171
Dasein: 183
deconstruction: 24
  "erasure": 49–50
  framework relativism: 24–25
  and logic: 49–50
  and metaphor: 185–86
  and new criticism: 3
  and origins: 177–80
  skepticism: 52–55. *See also* Derrida, Jacques, Nancy, Jean-Luc.
Deleuze, Gilles: 151–52n.2
demarcation, problem of: 65–66, 69–70, 74
Derrida, Jacques: 49–50, 52–55, 185–86
Descartes, René: 125
Devitt, Michael: 136n.16
Dickie, George: 223n.30
Dilworth, Craig: 93n.24, 101
discovery. *See* context of discovery, context of justification.
Douglas, Mary: 171n.25
dualism:
  methodological. *See* science/humanities dichotomy.
  ontological: 122, 125n.3
  and phenomenology: 129–32
Duhem, Pierre: 91, 102n.38
Dumouchel, Paul: 103n.40
Dupuy, Jean-Pierre: 49n.6, 256n.9
Durkheim, Emile: 78n.35, 115

Eckholm, Erik: 43n.4
economics: 167, 174–75, 261–62
Eddington, Sir Arthur: 85, 87
Einstein, Albert: 24, 27, 80n.2, 93–98
Eliade, Mircea: 51
Ellis, John: 159n.9, 235n.40, 253
emergence: 145–46
Empedocles: 1
empiricism: 43–44, 62
  basic tenets of, 90
epistemic access: 98–99. *See also* reference, truth.
epistemology: 38–42
  evolutionary: 134–37
  and literary theory: 16–18
  relation to "sociological turn": 65–79. *See also* knowledge, philosophy of science, skepticism, theory.
Escarpit, Roger: 214n.18
evidence: 43–44, 47, 60–64, 99, 115. *See also* justification, knowledge, skepticism, truth.
evolution:
  and epistemology: 134–37
  and literary criticism: 176
exactitude: 87–88
experimentation: 101–4, 115
explanation: 151–53, 225–32
  in literary criticism: 235–39
  versus understanding: 112, 239–40. *See also* causality.

falsification: 68
"family resemblance": 154–63
Faust, David: 66n.23
Favereau, Olivier: 174–75
feminism: 31n.28
Ferry, Luc: 26n.22
Feyerabend, Paul: 5, 25, 34, 80n.2
fiction. *See* literature, reference in.
Field, Hartry H.: 95–100
Fodor, Jerry A.: 189n.46
Foerster, Heinz von: 14n.8
formalism (literary critical): 215
forms of life: 163–69
Foucault, Michel: 70n.28, 150–52n.2, 190
foundations: 41–45, 130
framework relativism: 22–24, 33, 55–60, 115, 172–74
 critique of: 55–60, 172–74
 denial of scientific progress: 22–24
 plausibility of: 55–56
 and sociologism: 78–79
 theory of truth in: 56, 173
Frazer, Sir James G.: 168–70
Frege, Gottlieb: 163–164
Frongia, Guido: 155n.3
functionalism: 188–89

Galan, Frantisek: 265n.15
Galileo, Galilei: 64
Gardenförs, Peter: 64n.18
Gasché, Rodolphe: 3n.3, 24n.17
Gasking, Elizabeth B.: 134n.12
Genet, Jean: 190
Gibaldi, Joseph: 19
Giddens, Anthony: 113n.4
Gier, Nicholas F.: 166n.17
Girard, René: 105n.45
Glebe-Møller, Jens: 168n.19
Glymour, Clark: 227n.34
Gödel, Kurt: 80n.2
Goethe, Johann Wolfgang von: 219n.23
Goldmann, Lucien: 24
Gooding, David: 81n.3
Goodman, Nelson: 210n.9
Graff, Gerald: 196n.1, 197n.3
Gramsci, Antonio: 253
Granger, Gilles-Gaston: 104n.41
Gray, Bennison: 214n.17
Grayling, A. C.: 40n.2
Grimaldi, William: 212n.12
grounding. *See* foundations, justification, knowledge.
Grünbaum, Adolf: 113n.4

Habermas, Jürgen: 69, 109, 113, 191
Hacker, P.M.S.: 157, 162–63, 164–65
Hacking, Ian: 83n.5
Hall, Thomas S.: 134n.12
Hamlet: 139, 158–59, 253
Hanson, Norwood R.: 92–93, 100
Harré, Rom: 104
Harvey, William: 28, 105–6
Hegel, G.W.F.: 124, 191, 259
Heidegger, Martin: 27, 52n.11, 181–83
Hellman, Geoffrey: 84
Hempel, Carl G.: 226n.32
hermeneutics: 115, 132–34, 142
 and aesthetics: 222, 265–66
Hjort, Anne Mette: ix, 197n.3
Ho, Mae-Wan: 176n.35
Holub, Robert: 250n.3
Hudson, W. D.: 164n.15
humanities: 118–19
 experimentation in: 115
 idealistic assumptions: 132–34, 138, 145
 and naturalism: 122–23, 133
 relation to social sciences: 114. *See also* literary knowledge, science/humanities dichotomy.
Hume, David: 46, 168
Humphreys, Willard C.: 109n.52, 226n.32, 227n.36, 228–30, 243
Hunter, J.F.M.: 166n.17
Husserl, Edmund: 5, 9, 64, 85–86, 121–34, 138, 144, 146, 148, 157
Hutchinson, G. E.: 43n.4
hypothesis:
 and cognitive value of literary research: 260–63, 266–67

idealism: 84, 127–29, 138–39, 144–45, 148–49
idealization: 102, 205–6
idiography: 158. *See also* neo-Kantianism, science/humanities dichotomy.
idioi topoi/koinoi topoi: 211–212
incommensurability: 91–95, 100
Ingarden, Roman: 5, 138–45, 216, 221
Innis, George: 43n.4
inquiry. *See* explanation, humanities, knowledge, methodology, science, skepticism, theory.
institution: 16, 25, 40, 176, 216, 241–42. *See also* science.
instrumentalism: 86–87. *See also* constructivism, philosophy of science.
interdisciplinarity. *See* research, science.

interpretation. *See* hermeneutics, literary criticism, literary theory, central debate in.
Isis: 80n.1

Jacobi, F. H.: 174
Jakobson, Roman: 214
Jantsch, Erich: 120n.7
Jarry, Alfred: 152n.2
Jennings, Richard C.: 71n.29
justification (of claims to knowledge): 40–42, 47, 60–63, 104–7
 context of: 69–70, 73
 internalist theory of: 42

Kafka, Franz: 253
Kant, Imannuel: 27, 51, 70, 124, 127–28, 131, 132, 134, 140–41, 162n.13, 174, 213
Kepler, Johannes: 93, 100
Keynes, John Maynard: 175
Knapp, Steven: 9n.2
Knorr-Cetina, Karin D.: 69, 70n.28
knowledge:
 and authority: 58
 and literature: 2–3, 6
 pragmatic view of: 45, 57
 problem of: 35–36, 38–39
 and reflexivity: 184–85
 standard account of: 40–41
 value orientations: 30–31, 189–93. *See also* epistemology, justification, literary knowledge, philosophy of science, research, science, sociological turn, sociology of knowledge, theory, truth.
Konečni, Vladimir J.: 256n.10
Korzybski, Alfred: 207
Kristeller, Paul Oskar: 218n.22
Kuhn, Thomas: 5, 25, 35, 67, 69, 80n.2, 93–95, 96–101, 235n.40

Lacan, Jacques: 58
Lakatos, Imre: 80n.2
Lang, Candace: 26
language: 163–64
Laplace, Pierre Simon de: 108n.50
Laplacean demon:
 as image of science: 22, 118–19, 148, 228
Latour, Bruno: 67–68
Laudan, Larry: 71n.29
laws: 93n.24, 101, 225–28, 230–31
Leibniz, Gottfried Wilhelm von: 155
Leiris, Michel: 152n.2
levels of description: 88n.16, 145
Levin, Harry: 3n.2
Lévy-Bruhl, Lucien: 50–51
life-world (*Lebenswelt*): 85–86, 129, 131
List, Elisabeth: 168n.19
literariness. *See* literature, specificity of.
literary criticism:
 aesthetic a priori of: 216–24, 238–41
 aporia in: 243–44
 autonomy, lack of: 16–25, 254–65
 basic questions of: 233, 249, 250, 255, 260
 competence models: 202, 209–11
 cookie-cutter: 12
 crisis in: 209, 239–40
 evolutionary theory in: 176
 explanatory patterns of: 232–38
 focus on meaning vs. consequences: 249–52
 invisible theory in: 13
 methodological problems in: 232–33
 priorities of: 263–67
 sociological perspectives: 198, 244–58. *See also* aesthetics, deconstruction, literary knowledge, literary theory, "megaphone criticism."
literary knowledge: 196–99, 258–63
 relation to other models of knowledge: 16–25
literary theory:
 central debate in: 196–97, 200–4, 222
 images of science in: 22–29
 as prophecy: 11–12
 ways of construing: Chapter 1. *See also* aesthetics, epistemology, literary criticism.
literature:
 as attitude toward texts: 253
 autonomy of: 347
 definition of: 267
 oracular function: 234, 240, 265
 organicist doctrines of: 139–40, 219–22
 polysemy: 222–23
 reference in: 257–58
 specificity of: 214–18, 253–54, 264–65
 text/utterance distinction: 240
Livingston, Paisley: 223n.29, 262n.14
logic: 47–51, 64, 112
logical positivism. *See* positivism.
Luckmann, Thomas: 72
Lyotard, Jean-François: 56n.15

McCanles, Michael: 24–25
McCarthy, Thomas A.: 109n.51
McClelland, J. L.: 137n.17

McGuinness, B. F.: 157n.6
McIntyre, Ronald: 124n.2
Madsen, K. B.: 171n.24
Mallarmé, Stéphane: 27
Malpighi, Marcello: 105
Mandelbaum, Maurice: 145n.34
Mantzius, Karl: 258n.12
mapmaking, science as: 206–9
Marghescou, Mircea: 203n.1
Marmontel, Jean-François: 267
Martin, Raymond: 227n.35
marxism: 244, 249n.2, 255–56
materialism:
  reductionist: 86, 89–90, 145–46, 187–89. *See also* functionalism, ontology, philosophy of science, realism.
mathematics: 14–15, 85–86, 114
Maturana, Humberto: 132n.10
Mauss, Marcel: 170n.22
meaning:
  and knowledge: 210
  origin of: 177–81
  versus consequences: 245–52
"megaphone criticism": 233–34, 238–42, 265
Merleau-Ponty, Maurice: 126, 133n.11
Merton, Robert K.: 14n.6, 113
message: 234, 240, 264–65. *See also*: "megaphone criticism."
methodological dualism. *See* science/humanities dichotomy.
methodology: 12, 157–58, 171–72, 232–33
Meutsch, Dietrich: 223n.29
Michaels, Walter Benn: 9–10n.2
Michaut, Gustave: 258n.12
Millikan, Ruth Garrett: 88n.18, 101n.34
mind/body problem. *See* dualism, functionalism, materialism.
Moby Dick: 215
Molière: 2, 257–58, 260
Monod, Jacques: 134n.12
Moore, G. E.: 164, 168n.19
Morange, Michel: 136n.17
Morawetz, Thomas: 164n.15
Morawski, Stefan: 21, 266n.16
Moritz, Karl-Philipp: 218–22
Morris, Charles: 179n.37
Mulkay, Michael: 68n.25
Myers, Norman: 43n.4
mythology: 172–74, 229–31

Nadler, Joseph: 219n.23
Nancy, Jean-Luc: 26–28, 177–84
"natural attitude": 121–22. *See also* phenomenology.
nature: 110
  anthropomorphic views of: 78–79, 122–23, 128
neo-Kantianism: 111–12, 157, 162
Newton, K. M.: 241
Newton, Sir Isaac: 1, 93–98, 243
non-contradiction, principle of. *See* logic.
Novalis, 219n.23
Nowak, Leszek: 206n.3
Nozick, Robert: 40n.2, 95
Nussbaum, Martha: 175n.32
Nye, Mary Jo: 104n.42

Oakman, Robert: 215n.19
objectivity: 111–12n.1. *See also* justification, knowledge, realism, truth.
observation:
  direct: 43
  relation to theory: 44, 75–77, 82, 84–85, 87, 90–93, 102–7. *See also* incommensurability, realism, reference.
O'Connor, D. J.: 36n.2
O'Hear, Anthony: 168n.19
O'Neill, W. M.: 105n.46
ontology: 85–91, 137–38
  Seinsfrage: 182–83. *See also* dualism, idealism, materialism, realism.
organicism: 132, 139–40, 217, 219–22
origins:
  of meaning: 50, 176–83
  of sentience: 134–35
  skepticism about: *see* "always already" argument. *See also* deconstruction, framework relativism.
Orléan, André: 175n.30
Ort, Claus-Michael: 256n.10

paradigm (paradigm shift): 23, 25, 26n.21, 93–98. *See also* framework relativism, incommensurability, science, sociological turn.
paradox. *See* logic.
Pater, W. A. de: 212n.11
Peacocke, Arthur: 120n.7
Peirce, Charles Sanders: 84, 191
Perry, John: 257n.11
phenomenology:
  and anthropology: 121–23
  concept of "constitution": 125–27
  contemporary approaches to: 123–25
  critique of naturalist attitude: 128–29
  dualism of: 129–32
  and humanities: 132–34

phenomenology (*cont.*)
idealism: 127–28
transcendental subject: 128. *See also* Husserl, Edmund, Ingarden, Roman, "life-world."
Phillips, Derek L.: 157n.6
philosophy:
and disciplinarity: 211–14
postmodernist view of: 26–28
relation to science: 20. *See also* Heidegger, Martin, Wittgenstein, Ludwig.
philosophy of science:
basic positions in: 90–91
pertinence to literary research: 18, 20–28, 82. *See also* constructivism, demarcation, empiricism, explanation, materialism, neo-Kantianism, phenomenology, positivism, realism, science/humanities dichotomy, sociological turn.
Piaget, Jean: 191
Piché, Claude: 138n.10
Pippin, Robert B.: 36n.1
Plato: 40, 139, 155–56, 178
Poe, Edgar Allan: 21, 204
Ponech, Trevor: 92n.23
Popper, Sir Karl R.: 41, 90, 108n.50, 147
positivism: 65, 147
postmodernism: 56, 177–181
Pot-Bouille: 262
Pratt, Mary Louise: 217n.20, 257n.11
Prigogine, Ilya: 102–3, 120
progress:
distinction between types of: 35
in history: 191
in knowledge: 38, 127
postmodern critiques of: 23, 27–28
in science: 83–84, 90
psychoanalysis: 249n.1
Putnam, Hilary: 83n.5, 87, 100–1n.34

quantum mechanics: 80n.2, 105n.44
Quine, Willard Van Orman: 95–96

Rana paradoxa: 47–50
rationality of science: 101
Rausch, Hannelore: 11n.4
Raval, Suresh: 25, 161n.10
reading experiment: 201, 204–8
Reale, Giovanni: 213n.13
realism (epistemological):
basic tenets of: 34–35, 62–63, 83–84, 127
critique of constructivism: 106–7, Chapters 2 and 3
internal versus metaphysical: 87–88, 101n.34
minimal, moderate, and extreme: 88–90
non-reductionist: 92
reductionism. *See* materialism, reductionist.
reference (designation, denotation): 63–64, 77, 83, 91, 95–101, 185–88, 257
distinguished from re-presentation: 206–7. *See also* epistemic access, realism, truth.
regularities. *See* laws.
Reichenbach, Hans: 69n.26
relativism. *See* framework relativism.
Renaut, Alain: 26n.22
research:
as focus of Part Four: 199
interdisciplinary: 241, 259–63
and literary discipline: 2. *See also* knowledge, literary criticism, literary theory, science.
Rhees, Rush: 168n.19
rhetoric: 211–13
Richards, I. A.: 149
Richter, Jean Paul: 219n.23
Rickert, Heinrich: 112n.2
Ricoeur, Paul: 128n.7
ritual: 168–80
Rohatyn, Dennis: 49n.7
romanticism: 19, 21, 130, 218–21
Rorty, Richard: 57–58
Roussel, Raymond: 152n.2
Rudich, N.: 168n.19
Rudner, Richard S.: 14, 207
Rumelhart, D. E.: 137n.17
Russell, Bertrand: 46, 60, 163–64

Salmon, Wesley C.: 104n.42, 105, 227n.35
Saloman, Herman Prins: 258n.12
Saunders, Peter T.: 176n.35
Schaffer, E. S.: 152n.2
Schlegel, August Wilhelm: 219n.23
Schmidt, Siegfried J.: 29n.26, 223n.29, 233n.39, 254–55, 266n.16
Schnädelbach, Herbert: 20n.12, 112n.2
science:
disciplinarity: 211–13
distinguished from scientism: 107–10
as dominant model of knowledge: 20–22, 118–19
history of: 104–6
as Ideal: 28
impact on aesthetics: 21

science (*cont.*)
instrumental reliability of: 73–78, 104–7
as oriented map-making: 206–9
technology: 110
unity of: 147–48
value-orientations: 22, 29–31, 35, 66, 68, 107–10, 189–93. *See also* explanation, framework relativism, Laplacean demon, paradigm, philosophy of science, progress, rationality of science, science/humanities dichotomy, scientism.
science/humanities dichotomy: 1–2, 18, 111–20, 132–33, chap. 5.
approaches to: 114–19, 148–54
reflexivity argument: 184–86
typical positions: 111–14. *See also* "always already" argument, family resemblance, neo-Kantianism, phenomenology, positivism, reading experiment, romanticism.
scientism: 107–10
Scrivin, Michael: 227n.35
self-organizing systems: 136, 140, 191, 221–22
Sellars, Wilfrid: 85, 87
semantic indeterminacy: 95–100, 202, 240. *See also* meaning, message, reference.
semiology: 149, 209–10
Serres, Michel: 36n.2, 66, 88n.15
Shelley, Percy Bysshe: 19
Shope, Robert K.: 40n.2
Shusterman, Richard: 175–76
Sidney, Sir Philip: 196n.1
Siegel, Harvey: 69n.26, 101n.35
Simson, Rosalind S.: 40, 64n.19
skepticism: 44–49
coherent and incoherent kinds: 51–52, 55–56
Wittgenstein's criticism of: 164. *See also arrhepsia*, certainty, framework relativism, knowledge, problem of.
Smith, David Woodruff: 124n.2
Smith, Peter: 83–84
Smolensky, Paul: 137n.17
Sober, Elliott: 176n.35
sociological turn (in approaches to science): 65–79
sociology of knowledge: 66–68
arationality principle: 71
and epistemology: 70–79
Socrates: 155–56
Sommers, Fred: 64n.18
Sosa, Ernest: 40n.2
Spencer, Herbert: 176
Spengler, Oswald: 166
Spivak, Gayatri: 49–50
Sporn, Paul: 24
Stassen, M.: 168n.19
Stengers, Isabelle: 69, 102–3, 104n.43
Stockman, Norman: 65n.21, 83n.5
strong programme. *See* sociological turn.
Suppe, Frederick: 69, 84n.9

Taine, Hypolyte: 137, 149
Takei, Yushiro: 139n.19
Le Tartuffe: 256–58, 260
theoria: 10–11, 20, 129
theory: Chapter 1
epistemological role: 16–18
formal: 13–15, 163–64
invisible: 12
substantive: 10–13. *See also* literary theory.
Thomas, John: ix
Todorov, Tzvetan: 214n.18, 218–22
topics. *See idioi topoi/koinoi toooi.*
Toulmin, Steven: 69
truth:
consensus theory of: 39–40
framework relativist conception of: 56, 173
as goal: 190–93
as relation between belief and fact: 60, 62–63, 95–96. *See also* knowledge, reference.
Tugenhat, Ernst: 109n.51
Tuomela, Raimo: 84n.9, 85, 87–89, 105n.44, 251n.4
"two cultures." *See* science/humanities dichotomy.

understanding vs. explanation: 112, 239–40

value:
and human science: 191–94
idealist view of: 145n.33
and knowledge: 30, 35, 68, 108, 190–94
and priorities of literary criticism: 202, 257, 263–65
and social action: 30–31. *See also* science, value-orientations.
Varela, Francisco: 176n.35
Vattimo, Gianni: 80n.2
Vickers, Brian: 105n.46
Vollmer, Gerhard: 134–35n.13, 136

Warren, Austin: 3n.1, 139, 217n.21
Wartofsky, Marx W.: 88n.15, 206–8
Watzlawick, Paul: 81n.4
Weber, Max: 206n.3
Weitz, Morris: 158–59
Wellek, René: 2–3, 139, 214, 216–18, 221–22
Weltbild (world picture): 172
Whitehead, Alfred North: 108
Whitley, Richard: 67–68, 70n.28
Williams, David: ix
Winfree, Arthur T.: 120n.7
Wittgenstein, Ludwig: 5, 27, 87n.15, 92, 154–76, 244
Woolgar, Steven: 67–68
Wright, G. H. von: 164n.15, 251n.4
writing: 104
Wylie, Alison: 105n.47

Znaniecki, Florian: 67n.24
Zola, Emile: 262
Zumbach, Clark: 132n.10

Library of Congress Cataloging-in-Publication Data

Livingston, Paisley, 1951–
Literary knowledge.

Includes index.
1. Literature—Philosophy. 2. Knowledge, Theory of. 3. Literature and science. 4. Science—Philosophy. I. Title. II. Title: Humanistic inquiry and the philosophy of science.
PN49.L54 1988 801'.9 87-47821
ISBN 0-8014-2110-1 (alk. paper)
ISBN 0-8014-9422-2 (pbk. : alk. paper)